Framing a Strategic Approach for

Joint Officer Management

Harry J. Thie, Margaret C. Harrell, Roland J. Yardley,
Marian Oshiro, Holly Ann Potter,
Peter Schirmer, Nelson Lim

Prepared for the Office of the Secretary of Defense
Approved for public release; distribution unlimited

NATIONAL DEFENSE RESEARCH INSTITUTE

The research described in this report was prepared for the Office of the Secretary of Defense (OSD). The research was conducted by the RAND National Defense Research Institute, a federally funded research and development center supported by the OSD, the Joint Staff, the unified commands, and the defense agencies under Contract DASW01-01-C-0004.

Library of Congress Cataloging-in-Publication Data

Framing a strategic approach for joint officer management / Harry J. Thie ... [et al.].
    p. cm.
    "MG-306."
    Includes bibliographical references.
    ISBN 0-8330-3772-2 (pbk. : alk. paper)
    1. United States—Armed Forces—Officers. 2. United States—Armed Forces—
Personnel management. 3. Unified operations (Military science) I. Thie, Harry.

UB413.F73 2005
355.3'32—dc22

2005006347

The RAND Corporation is a nonprofit research organization providing objective analysis and effective solutions that address the challenges facing the public and private sectors around the world. RAND's publications do not necessarily reflect the opinions of its research clients and sponsors.

RAND® is a registered trademark.

Cover design by Stephen Bloodsworth

Published 2005 by the RAND Corporation
1776 Main Street, P.O. Box 2138, Santa Monica, CA 90407-2138
1200 South Hayes Street, Arlington, VA 22202-5050
201 North Craig Street, Suite 202, Pittsburgh, PA 15213-1516
RAND URL: http://www.rand.org/
To order RAND documents or to obtain additional information, contact
Distribution Services: Telephone: (310) 451-7002;
Fax: (310) 451-6915; Email: order@rand.org

Preface

The National Defense Authorization Act for Fiscal Year 2002 directed an independent study of joint officer management, joint professional military education (JPME), and the roles of the Secretary of Defense and the Chairman of the Joint Chiefs of Staff. Started in September 2002, the independent study sought to determine the effectiveness of joint officer management and JPME Phase II based on the implications of proposed joint organizational and operational concepts (such as standing joint force headquarters) and emerging officer management and personnel reforms being considered by the Secretary of Defense.

While the independent study was in progress, the General Accounting Office (GAO)[1] conducted an assessment of Department of Defense (DoD) actions to implement provisions in the law that address the development of officers in joint matters and evaluated impediments affecting DoD's ability to fully respond to the provisions of the Goldwater-Nichols Act (GNA) of 1986. The report also addressed the challenges DoD experiences in preparing officers to serve in joint organizations and leadership positions in terms of education, assignment, and promotion. The GAO recommended that DoD develop a strategic approach to the development of officers in joint matters. Such a strategic human resource management approach would establish clear goals for officer development in joint matters

---

[1] Renamed the Government Accountability Office in July 2004.

and would link joint officer development to DoD's overall missions and goals.

In March 2003, the independent study was completed and indicated that joint officer management/JPME II requires updating in practice, policy, and law to meet the demands of a new era more effectively. The study concluded that change is warranted to develop better the officer corps for joint warfare and that change should be undertaken as part of an overall strategic approach to developing the officer corps for joint warfare.

The RAND National Defense Research Institute, a division of the RAND Corporation, was asked to undertake an analysis that is intentionally broad, looking beyond joint manpower issues to establish the context for officer development in joint matters. This analysis is designed to conceptualize a strategic approach for officer development in joint matters. The intent of a strategic approach is to provide overarching guidance on officer training and development in joint matters to best meet DoD's mission and goals in the context of evolving combatant commander and service requirements, revolutionary changes in technology, and a dramatic cultural shift in the military.

This report communicates the findings of our joint officer analysis, our conceptual strategic approach for joint officer management, and our recommendations for operationalizing the strategic plan. This report is intended for those both interested and informed in joint personnel matters. Although it does include an appendix that provides a basic introduction of GNA requirements and joint officer management issues (Appendix A), this report is not intended for those wholly unfamiliar with joint officer management matters but instead follows and responds to the GAO report on this topic.

This research was sponsored by the Under Secretary of Defense for Personnel and Readiness. It was conducted within the Forces and Resources Policy Center of the RAND National Defense Research Institute, a federally funded research and development center sponsored by the Office of the Secretary of Defense, the Joint Staff, the unified commands, and the defense agencies. The principal investigators are Harry Thie and Margaret Harrell. Comments are welcome and may be addressed to Harry Thie at harry_thie@rand.org or to Margaret Harrell at margaret_harrell@rand.org.

For more information on RAND's Forces and Resources Policy Center, contact the director, Susan Everingham. She can be reached by email at susan_everingham@rand.org; by phone at 310-393-0411, extension 7654; or by mail at the RAND Corporation, 1776 Main St., Santa Monica, CA 90407-2138. More information about RAND is available at www.rand.org.

Contents

Figures and Tables

Figures

Tables

Summary

Background and Purpose

The Goldwater-Nichols Act (GNA) of 1986 forged a cultural revolution in the U.S. armed forces by improving the way the Department of Defense (DoD) prepares for and executes its mission.[1] Title IV of the GNA addresses joint officer personnel policies and provides specific personnel management requirements for the identification, education, training, promotion, and assignment of officers to joint duties.

Recent studies[2] suggest the need for DoD to revisit joint manpower matters and develop a strategic approach to joint officer management and joint professional military education (JPME).[3] Additionally, the National Defense Authorization Act for Fiscal Year 2002 directed an independent study of joint officer management, JPME, and the roles of the Secretary of Defense and the Chairman of the Joint Chiefs of Staff. While the independent study was in progress, the General Accounting Office (GAO)[4] conducted an assessment of DoD actions to implement provisions in law that address the devel-

---

[1] Goldwater-Nichols is discussed in more detail in Appendix A, which provides a primer for many of the terms and concepts discussed herein.

[2] General Accounting Office, *Joint Officer Development Has Improved, but a Strategic Approach Is Needed*, GAO-03-238, 2002; Booz Allen Hamilton, *Independent Study of Joint Officer Management and Joint Professional Military Education*, McLean, Va., 2003.

[3] Joint Professional Military Education is discussed in more detail in Appendix A.

[4] Renamed the Government Accountability Office in July 2004.

opment of officers in joint matters. It also evaluated DoD's ability to fully respond to the provisions of the GNA.

The GAO stated that "a significant impediment affecting DoD's ability to fully realize the cultural change that was envisioned by the act is the fact that DoD has not taken a strategic approach to develop officers in joint matters."[5]

A strategic approach to human resource management determines the need for critical workforce characteristic(s) given missions, goals, and desired organizational outcomes; assesses availability of the characteristic(s) now and in the future; and suggests changes in management practices for personnel with the characteristic(s) to minimize gaps between need and availability. This report applies a strategic approach to the development of officers in joint matters.

Research Approach

This strategic approach was developed consistent with the human resources literature regarding the purposes, intents, and qualities of strategic approaches. To assess the amount of joint experience or joint education currently available among the officer corps, we conducted detailed statistical analysis of longitudinal data files constructed from the officer master file. The quantitative analysis that we prescribe will support a determination of the need for and the provision of desired work characteristics (e.g., to what extent different positions either require or provide joint experience). In addition, while developing this strategic approach, we conducted interviews and group discussions with many officers to gain insights regarding the characteristics that positions would be likely to require (or provide), problems and shortcomings of the existing system, likely resistance to or difficulties in developing a new management system, and other helpful suggestions regarding a new management system for joint officers. Officers from each of the services' officer management offices participated in

---

[5] General Accounting Office, *Joint Officer Development Has Improved, but a Strategic Approach Is Needed*, GAO-03-238, 2002

these discussions, as did officers from each of the senior service schools. We also traveled to several combatant commands (EUCOM, PACOM, SOCOM), where we interviewed officers in leadership positions and conducted group discussions with officers in pay grades O-4 through O-6 from their J-1, J-2, J-3, J-4, J-5, and J-6 offices. We also conducted similar group discussions with officers from each of the collocated service component commands.

Management of the Joint Duty Assignment List, and Joint Officer Development

The President's National Security Strategy as well as service and joint vision documents describes increasingly joint missions, but officer management is following the trend more slowly. Data indicate a decreasing fill rate of joint duty assignments (JDAs) for three of the four services. (The Marine Corps is not only increasing the rate at which it fills joint assignments, but also increasing its share of the Joint Duty Assignment List [JDAL].) The service personnel managers (for all but the Marine Corps) note the difficulty in fitting joint assignments into officers' career paths and are reflective of individual service cultures that are generally less respectful of joint experience than of that gained within their services. Nonetheless, longitudinal data indicate an increasing amount of jointness among the officer corps.

Four general observations emerge from our look at these data.

- In general, the data give evidence that jointness is accreting in the officer corps. Officers as a group are more "joint" with each passing year. Officers who are joint specialty officers (JSOs), who are joint qualified, or who have some joint experience advance in grade and stay in service at rates sufficient to have increased overall joint content of the officer corps over time. This is more accurately stated for certain grades, occupations, and services than for others.

- One can draw different inferences by looking at the data in different ways. For example, if one examines the entire officer corps for grades O-4 and above, the fact that it has accumulated about 15 percent of officers with at least some joint experience does not seem high.[6] However, the denominator in the equation consists of many officers in occupations not inherently joint (e.g., health) and is more heavily weighted toward O-4s who have had less time to be joint. As one focuses on certain occupations (e.g., intelligence, tactical operations) or on particular grades in which a higher cumulative opportunity to become joint has existed (e.g., O-6), one sees a far rosier picture of the accretion of jointness in the officer corps.
- Increases in jointness have consistently occurred each year. However, a cursory look at the data indicates that such increases are leveling off, i.e., becoming asymptotic at current levels. How joint the officer corps can be is dependent on the opportunity to have a JDA and to attend JPME Phase II. The seats for the latter are limited, and the number of the former is also limited. It may be that, given these constraints, underlying job and educational durations, and continuation and promotion rates, the ability to increase jointness further in the officer corps in the future may not exist absent changes in the number and duration of school and assignment seats. These limits can be explored with career models of the type we discuss later in this report.
- While there are observable differences in behaviors and outcomes between those with and without joint experience, these differences are more apparent in data describing officers leaving the service than in data showing officers' advancement in their services, and many of these differences either may not be signifi-

---

[6] We parsed joint experience into three groups. The first are those who have received full credit for a joint tour but are not JSOs. The second are those who are JSOs. The third are those who at a point in time are not in the first two groups but have some joint experience in a qualifying position. For example, they may be currently serving in such a position or they may have received partial credit for past service. We also track officers who have no joint experience in a qualifying JDAL position. We will differentiate among these groups for some of the figures and analysis in Chapter Three.

cant or may result from other factors than jointness. For example, promoting more to O-4 in one year than in other years can change the joint content at that grade for one or more years.

A Strategic Approach to Joint Officer Management

A strategic approach must understand the need or requirement for critical workforce characteristics and the ability of the management system to provide officers with those characteristics. Moreover, the approach needs to demonstrate (1) a strategy or policy for aligning the availability of officers with the characteristic with the need for them or (2) a rationale for why more widespread availability of the characteristic than the immediate need for it would be desirable. A strategic approach for joint officer management must deliberately determine which jobs, inside or outside the service, require joint experience or provide it. In particular, given the current strategic intent of DoD with respect to jointness ("push it to its lowest appropriate level"), the need for joint experience should be measurable in a much larger number of billets—in particular, in billets internal to the service. Moreover, valid joint experience might now be provided by service in billets internal to the service, particularly those associated with joint task forces, with service component commands, and with joint planning and operations. The key components of a strategic approach can be discerned as (1) which jobs require or provide joint experience, (2) how many of each exist, and (3) what is needed to align those two sets of jobs.

Implementing a Strategic Approach for Joint Officer Management

Our recommended approach has five major steps:

1. Define workforce characteristics that will be needed in the future to meet strategic intent. We believe that these characteristics can

be aggregated into proxy variables for competencies based on experiences such as joint multiservice, joint interagency, and joint multinational[7] and on joint education and/or joint training. The accuracy of billet needs with respect to characteristics such as grade (experience), occupation, and other characteristics will need to be assumed.

2. Define needs for these characteristics of joint experience, education, and training. Where (in what positions) are officers with joint experience, education, and training needed? How many of these positions are there? Does this differ across services, for different occupations, or at different levels of seniority? Does the need for such officers extend to in-service billets? For this step, we recommend collecting data as to which external and in-service billets need an officer with prior joint experience or education.

3. Identify officers with these characteristics who are currently available. We recommend using existing personnel databases to assess the current numbers of officers with the experience and education characteristic of interest. We recommend surveying all external billets (and selected in-service billets) to determine those billets that provide joint experience to officers as a basis for projecting future availability. Current numbers and timing of JPME II seats[8] will likely need to be used as the start point for projecting educational qualifications.

4. Use models to

   a. Project availability of officers with these characteristics in the future, given certain career management practices

---

[7] We have begun to see use of the acronym "JIM" to reflect joint (multiservice), interagency, and multinational as separate components of the larger concept of jointness. This will be discussed further in Chapter Five.

[8] As of this writing, the House Armed Services Committee has a provision in its version of the 2006 NDAA that would increase the number of institutions that could provide JPME II and thus increase the number of JPME II graduate rates. Depending on the final outcome of this provision, we will incorporate any changes in the modeling approach we take in the next phase of our research.

b. Calculate future gaps between the need for officers and the availability of them[9]
c. Refine and evaluate near-term policy alternatives to reduce gaps within the strategic context
d. Develop strategies that address long-term issues for reducing the gaps.
5. Identify other implications of the strategic approach such as effects on objectives and desired metrics for evaluation.

In summary, the strategic approach needs to confirm the characteristics of interest. We assert they are likely multiservice, multinational, and interagency experience as well as joint training and joint education. Data need to be gathered that will confirm the characteristics of interest as well as quantify the need for jointness, the current stock of jointness, and the possible future provision of jointness. This report contains protocols and instructions for collection both types of data. Modeling should confirm the degree to which jointness can be accumulated and the extent to which the future stock of jointness will satisfy the future identified demand for jointness.

Identifying Policy Implications

Using the data on the need for and availability of the workforce characteristics and the management frameworks, the analysis to implement the strategic approach should provide input to such policy issues as the following:

---

[9] We are using the logic that underlies strategic human capital management of matching availability of workforce characteristics to the demand for them. This assumes that there is a cost for developing people with these characteristics so that both an over and under supply of the characteristic is not desirable. However, other assumptions could be made that change the nature of the assessments we are making. For example, the availability of officers with joint experience and/or education could lead to increasing demand for them in many military positions. The availability of such officers could, by itself, create a need for them.

- Which variables should be included in a definition of qualifying joint experience?
- How "joint" should a billet be in order to be considered validly joint?
- Should joint experience gained from multiservice, multinational, and interagency billets be managed or tracked separately? Are the needs and sources of each distinctly separable?
- Should minimizing oversight and repetitive measurement be a consideration? For example, if an organization is 95 percent joint, should all billets at that organization receive joint credit? If all billets are somewhat joint, should all billets provide joint credit?
- What management frameworks are suggested for different services and occupations? How different is that from the status quo?
- Are there other occupational considerations? For example, will some career fields have more difficulty gaining jointness?
- What is the relationship between necessary education and training and existing resources? Are more resources (seats) needed?
- What should be the objectives and metrics of a system to develop officers in joint matters?

Conclusions and Recommendations

Goldwater-Nichols deserves some reconsideration, given the increasing number of joint operations, the recognition of the value of jointness among officers, and the changing management practices for officers with joint experience. All the original objectives of the GNA may not still be appropriate, and considerable conflict exists within the GNA objectives as well as between the GNA objectives and the stated goals of the services, the joint organizations, and individual officers.

However, it is not clear that the types of constraints and requirements stated in the GNA should be eliminated. Military missions are increasingly integrated, and military officers are increasingly joint. However, there still exists some cultural resistance to officers' jointness. In other words, the officer management systems in most of the services are still generally resistant to developing joint officers and would likely revert to management processes that did not support jointness in the absence of GNA-type requirements, constraints, and reporting mechanisms.

A strategic approach to joint officer management, as outlined here, aligns human capital with the organization's mission, rather than empowering other influences, such as organizational, administrative, and cultural heritage or the current social, cultural, and legal practices and beliefs. The strategic approach described herein for joint officer management considers and balances the assignments that require joint experience, education, training, or acculturation with the ways officers receive joint experience, education, training, or acculturation.

The next research step is to operationalize, or implement, the strategic plan for joint officer management. This implementation will require extensive data gathering and complex modeling and data analysis in order to formulate appropriate policy alternatives. This report provides the implementable means to do this.

Acknowledgments

This research benefited from the many individuals in combatant commands, service component commands, and on service staffs who agreed to speak candidly with us regarding joint personnel management and service policy and attitudes regarding joint issues. We are grateful for the support and interaction of Major Harvey Johnson, Lieutenant Colonel Charles Armentrout, Brad Loo, and Gwendolyn Rutherford from our sponsoring office and the input provided by Lt Col Timothy Nelson, CDR Carolyn Owens, Lt Col Charlene Jefferson, and Karen Miller of JCS J1.

We are indebted to our reviewers, RAND colleagues John Boon, John Schank, and Mike Thirtle. We also appreciate the contributions of colleagues Matt Schonlau, Terry West, and Marc Elliott and the administrative support provided by Sonia Nagda and Samantha Merck. Also at RAND, we thank graphic designer and artist Stephen Bloodsworth and editor Phillip Wirtz for their assistance with the final version of the report.

Abbreviations

AFS	air force specialty
AFSC	air force specialty code
AOR	area of responsibility
AQD	additional qualification designation
CJCS	Chairman of the Joint Chiefs of Staff
CJTF	combined joint task force
COMPACFLT	Commander, Pacific Fleet
COMUSNAVSO	Commander, U.S. Naval Forces, Southern Command
COS	critical occupation specialty
CUSNFS	Commander, U.S. Naval Forces
DoD	Department of Defense
EUCOM	European Command
GAO	General Accounting Office [renamed the Government Accountability Office in July 2004]
GNA	Goldwater-Nichols Act
IA	interagency
ICAF	Industrial College of the Armed Forces
JDA	joint duty assignment
JDAL	Joint Duty Assignment List

JFSC	Joint Forces Staff College
JOM	joint officer management
JOPES	Joint Operation Planning and Execution System
JPME	joint professional military education
JSO	joint specialty officer
JTF	joint task force
MN	multinational
MOS	military occupational specialty
MS	multiservice
NDAA	National Defense Authorization Act
NSS	National Security Strategy
NWC	National War College
PACOM	Pacific Command
POC	point of contact
SEI	special experience identifiers
SOCOM	Special Operations Command
SOF	special operation forces
UCC	unified combatant commander

CHAPTER ONE

Introduction

Background

The Goldwater-Nichols Act (GNA) of 1986 forged a cultural revolution in the U.S. armed forces by improving the way the Department of Defense (DoD) prepares for and executes its mission.[1] Title IV of the GNA addresses joint officer personnel policies and provides specific personnel management requirements for the identification, education, training, promotion, and assignment of officers to joint duties. The summarized objectives of Title IV are as follows:

- Enhance joint warfighting capabilities.
- Increase the quality of officers in joint assignments.
- Ensure that officers are not disadvantaged by joint service.
- Ensure that general and flag officers are well rounded in joint matters.
- Enhance the stability and increase the joint experience of officers in joint assignments.
- Enhance the education of officers in joint matters and strengthen the focus of professional military education in preparing officers for joint duty assignments (JDAs).

---

[1] Goldwater-Nichols is discussed in more detail in Appendix A, which provides a primer for many of the terms and concepts discussed herein.

In the past 17 years, successes in Iraq (Desert Shield/Storm), Bosnia, and Afghanistan (among others) and most recently in Operation Iraqi Freedom testify to the progress and effectiveness of the joint military force and its warfighting potential. However, recent studies[2] point to the need for DoD to revisit joint manpower matters and develop a strategic approach to joint officer management and joint professional military education (JPME). A strategic approach would provide overarching guidance on how officer training and development in joint matters would best meet DoD's mission and goals in the context of evolving combatant commander and other personnel requirements, considerably reduced service resources, revolutionary changes in technology, and a dramatic cultural shift in the military.

The National Defense Authorization Act (NDAA) for Fiscal Year 2002 directed an independent study of joint officer management, JPME, and the roles of the Secretary of Defense and the Chairman of the Joint Chiefs of Staff (CJCS). While the independent study was in progress, the General Accounting Office (GAO)[3] conducted an assessment of DoD actions to implement provisions in law that address the development of officers in joint matters. It also evaluated DoD's ability to fully respond to the provisions of the GNA:

- Education: Although a two-phased joint education program has been developed, DoD has not determined how many officers should complete both phases.
- Assignment: Critical joint duty positions have not been filled with officers meeting the prerequisites mandated for these positions.
- Promotion: Although DoD has promoted more officers with joint experience to flag and general officer positions, the

---

[2] General Accounting Office, *Joint Officer Development Has Improved, but a Strategic Approach Is Needed,* GAO-03-238 2002; Booz Allen Hamilton, *Independent Study of Joint Officer Management and Joint Professional Military Education,* McLean, Va., 2003.

[3] Renamed the Government Accountability Office in July 2004.

department still relies on waivers for joint experience to promote officers to these senior levels. Further, although DoD has promoted mid-grade officers with joint experience at a rate similar to or better than their peers, the department is challenged with meeting this goal for colonels and Navy captains.

The GAO stated that "a significant impediment affecting DoD's ability to fully realize the cultural change that was envisioned by the act is the fact that DoD has not taken a strategic approach to develop officers in joint matters." The GAO recommended that DoD develop a strategic approach that will establish clear goals for officer development in joint matters and link joint officer development to DoD's overall missions and goals. Further, the strategic approach should

- Identify the number of joint specialty officers (JSOs) needed
- Provide for the education and assignment of reservists who may serve in joint organizations
- Be developed to provide more meaningful data to track progress made against the plan.

The independent study concluded that change is warranted to better develop the officer corps for joint warfare, that change should be undertaken as part of an overall strategic approach to developing the officer corps for joint warfare, and that such change should be led by the Secretary of Defense and the CJCS.

DoD's Mission and Goals Express Increasing Jointness

Since the Goldwater-Nichols, the imperative to conduct joint operations to maximize the capabilities of the force has been recognized and incorporated into strategic plans, vision and mission documents, from the National Security Strategy (NSS) and National Military

Strategy (NMS) through the planning guidance of the individual services.[4] These documents serve to identify DoD's overall missions and goals for joint officer management. Although not specifically stated, the strategic intent as it relates to joint officer development can be derived from the NSS as defending our nation against its enemies through the strengthening of joint operations. The NMS calls for U.S. armed forces to be multi-mission capable, interoperable among all elements of U.S. services and selected foreign militaries, and able to coordinate operations with other agencies of government and some civil institutions.[5] Further, the overarching guidance in the Quadrennial Defense Review (2001), the Secretary of Defense's Transformation Planning Guidance (2003), and Joint Operations Concept (2003) calls for future military responses to be joint and that DoD must transform to meet the challenges of the 21st century. Finally, future vision documents (Joint Visions 2010 and 2020) state that the joint force remains key to operational success in the future. To be effective the force must be "intellectually, operationally, organizationally, doctrinally, and technically joint."[6]

The services have recognized these goals and the imperative to operate as a joint team to execute DoD's mission. In their most recent vision statements:

- The Army describes how it is exploring what it must become to provide more relevant and ready forces and capabilities to the joint team.
- The Navy vows to work together with its sister services as a "joint" team, committed to and built on the principles of jointness.
- The Air Force states a goal of domination of the aerospace domain to facilitate the effectiveness of the joint team.

---

[4] This is discussed in more detail in Chapter Two.

[5] Chairman of the Joint Chiefs of Staff, *National Military Strategy: Shape, Respond, Prepare Now: A Military Strategy for a New Era*, 1997.

[6] Chairman of the Joint Chiefs of Staff, *Joint Vision 2020*, 2000.

- The Marine Corps undertakes to reinforce its strategic partnerships with sister services and contribute to the development of joint, allied, coalition, and interagency capabilities.

Joint Officer Management Is Not as Far Along as Joint Missions

As discussed previously, DoD leadership, including the service chiefs, has enunciated strong views with respect to jointness of strategy, plans, and operations, especially for the future. However, even while increasingly joint, officer management is following this trend more slowly, and the military culture, while supportive of joint missions, may still be resistant to developing joint officers. Cultures progress from fragmentation to differentiation to integration.[7] For most of their histories, the military services can be described as fragmented. Each brought unique capabilities and each desired to use these capabilities independently of the other services. More recently, the services could be described as differentiated in that the capabilities each brought were distinct but could operate jointly (interoperable) with the capabilities of the other services. Most recently, arguments are made that the capabilities of the services are integrated and represent a whole rather than the sum of parts. On the operational side, claims are plausibly made that integration of the services[8] has started and that further integration is accelerating. Such claims are not as plausible on the management side. Joint officer management does not appear to be as far advanced.

Locher's statement in 1996[9] that DoD "lacks a vision for its needs for joint officers and how to prepare and reward them" still

---

[7] These terms are drawn from Joann Martin, *Organizational Culture: Mapping the Terrain,* Thousand Oaks, Calif.: Sage, 2002.

[8] This section focuses on multiservice integration of capability. Similar arguments must be made for interagency and multinational capability.

[9] James R. Locher III, "Taking Stock of Goldwater-Nichols," *Joint Forces Quarterly,* Autumn 1996.

accurately describes the current state of the system. The CJCS, in recent speeches and testimony,[10] has also expressed views about the state of the current system:

- "Joint officer management must evolve to reflect the way we operate in today's environment."
- "We must create a military culture that embraces a new level of collaboration between the services."
- "Doctrine, organization, and training must be focused not only on developing service expertise but also on creating experts in melding service capabilities."

A Strategic Approach

A strategic approach could resolve the disparity between the increasing jointness of military missions and the less-joint officer management system.[11] A strategic approach must understand the need or requirement for critical workforce characteristics as a result of missions and goals and the ability of the management system to provide officers with those characteristics. Moreover, the approach needs to demonstrate (1) a strategy or policy for aligning the availability of officers with the characteristics with the need for them or (2) a rationale for why more widespread availability of the characteristic than the immediate need for it would be desirable.

---

[10] Chairman of the Joint Chiefs of Staff, "Posture Statement," February 8, 2002; 31st Annual IFPA [Institute for Foreign Policy Analysis]–Fletcher Security Conference, Washington, D.C., November 14–15, 2001; *Joint Forces Quarterly*, Autumn/Winter 2001–2002.

[11] The Secretary of Defense has questioned whether the right officers are getting into joint assignments and into combatant commands, whether the positions that now give joint service credit are really very joint, and whether positions that are currently not getting joint credit should. Moreover, our interviews with combatant commands, service component commands, and the military services raised these and other management issues.

Conflicting Objectives Are an Impediment to a Strategic Approach

Goldwater-Nichols establishes a set of objectives for joint officer management. Our interviews with service representatives and with officers in leadership roles in joint organizations, as well as previous research, suggest additional objectives. The summarized objectives are listed below:[12]

- Ensure that military leaders have joint experience.
- Provide officers with prior joint experience and education to joint organizations.
- Ensure that joint organizations receive quality officers.
- Ensure that officers are not disadvantaged because of joint service.
- Maximize the number of officers with joint experience; increase joint understanding within the services.
- Minimize disruption to service careers; minimize the time away from service for everyone.
- Minimize the time away from service for the best officers.
- Reduce organization instability by increasing tour length and thus also increase individual accountability.
- Provide officers with deep, current service expertise to joint organizations.
- Provide officers with credit for a JDA and JPME Phase II.

Some of these objectives gathered through interviews and discussions conflict with the objectives of the GNA. For instance, it is difficult to minimize the time away from the service for the best officers while ensuring that tomorrow's leaders are well rounded in joint matters. Likewise, minimizing the time away from the service for best

---

[12] Additional interviews or different interpretations of the GNA may produce additional objectives.

officers is not consistent with the longer assignments suggested by the GNA objective to enhance the stability and accountability in organizations. Additionally, if the GNA wants to increase or maximize the number of officers serving in joint assignments who have prior joint experience, it may be difficult to satisfy the need of joint organizations for officers with rich, current service expertise.

Other conflicts exist within the list of GNA objectives itself. Given a limited number of joint assignments, if the intent is to ensure that future general and flag officers are well rounded in joint assignments, then it may not be possible to also maximize the number of officers who gain joint experience; those assignments may need to be targeted toward future leaders. Additionally, if the GNA would prioritize assignment of the best quality officers (future leaders) to joint assignments, then stabilizing or lengthening these assignments may not be an obtainable objective.

In sum, the GNA establishes certain objectives for joint officer management. The services, joint organizations, and the officers themselves also offer a set of objectives. There are conflicts not only between GNA objectives and those offered within DoD, but also between GNA objectives themselves. These objective conflicts suggest a system with internal conflicts, or one that is out of sync with its context, or both. Not only do proponents of jointness have different perspectives and objectives than service proponents, but there are conflicting objectives within the joint community. These disconnects suggest yet another justification for reexamining joint officer management. It is possible that some objectives are no longer valid—that they were important during the transitional period of establishing a strong joint community but are less important now. Alternatively, it is possible that while some objectives should apply to the entire DoD, others will apply differentially to different populations of officers. This will become increasingly clear in the later discussion of different management frameworks, where different frameworks support different objectives, applicable to different groups of officers.

Purpose of the Report

This report provides a framework and an executable process for implementing a strategic approach to joint officer management. This is intended to supplement prior work that established the need for such an approach.[13] This work also builds on prior RAND research that quantified the extent to which the services could develop officers with joint education and joint experience and explored how to determine the level or amount of joint experience provided by different billets.[14] This prior work, however, assumed that the need for jointness was limited to select JDAs, those already established as critical positions thus requiring JSOs. Instead, this strategic approach suggests that the extent to which billets provide joint experience or require prior jointness (either education or experience) will not necessarily correlate with whether they are currently on the JDAL or whether they reside in joint organizations.

Research Approach

This strategic approach was developed consistent with the human resource literature regarding the purposes, intents, and qualities of strategic approaches. To assess the amount of joint experience or joint education currently available among the officer corps, we conducted detailed statistical analysis of longitudinal data files constructed from the officer master file.[15] The quantitative analysis that we prescribe will support a determination of the need for as well as the provision

---

[13] Booz Allen Hamilton (2003); General Accounting Office (2002).

[14] Harrell, Margaret C., John F. Schank, Harry J. Thie, Clifford M. Graff II, and Paul Steinberg, *How Many Can Be Joint? Supporting Joint Duty Assignments,* Santa Monica, Calif.: RAND Corporation, MR-593-JS, 1996; John F. Schank, Harry J. Thie, Jennifer Kawata, Margaret C. Harrell, Clifford M. Graf II, and Paul Steinberg, *Who Is Joint? Reevaluating the Joint Duty Assignment List,* Santa Monica, Calif.: RAND Corporation, MR-574-JS, 1996; John F. Schank, Harry J. Thie, Margaret C. Harrell, *Identifying and Supporting Joint Duty Assignments: Executive Summary,* Santa Monica, Calif.: RAND Corporation, MR-622-JS, 1996.

[15] This analysis is presented in Chapter Three.

of desired work characteristics (e.g., to what extent different positions either require or provide joint experience). In addition, while developing this strategic approach, we conducted interviews and group discussions with many officers to gain insights regarding the characteristics that positions would be likely to require (or provide), problems and shortcomings of the existing system, likely resistance to or difficulties in developing a new management system, and other helpful suggestions regarding a new management system for joint officers. Officers from each of the services' officer management offices participated in these discussions, as did officers from each of the senior service schools. We also traveled to several combatant commands (EUCOM, PACOM, SOCOM), where we interviewed officers in leadership positions and conducted group discussions with officers in pay grades O-4 through O-6 from their J-1 through J-6 offices. We also conducted similar group discussions with officers from each of the collocated service component commands.

Organization of the Report

This research was intentionally broad, looking beyond joint manpower issues to establish the context for officer development in joint matters. Thus, the research documented in this report was designed to establish a strategic approach for officer development in joint matters and guide the follow-on research to operationalize[16] the strategic approach for joint officer issues.

This introduction contains background material regarding the increasing jointness of our military missions the need for a strategic approach to joint officer management. It discusses issues related to the GNA, including the conflicting objectives of joint officer management, and describes our research approach. This chapter is sup-

---

[16] The authors acknowledge the liberty we have taken by using the word "operationalize." We chose this word to underscore that making a strategic approach for joint officer management operational is more involved than simple implementation, as a strategic plan can involve some significant changes from the status quo.

plemented by Appendix A, a primer to the GNA. Chapter Two establishes the need for joint officers in current and future joint missions.

Chapter Three provides the empirical baseline of how the services have managed JDAs and how officers with joint duty experience have been managed. This latter discussion is based on data that indicate the "jointness" of the officer force and the extent to which joint officers have retained and been promoted over their careers. This chapter also provides a description of the current service attitudes toward joint officer management and development. This chapter is important to a strategic approach in that it identifies the current availability of joint characteristics.

Chapter Four discusses the characteristics of and need for a strategic approach, and Chapter Five details the application of a strategic approach to joint officer management. This chapter provides considerable detail necessary to follow-on efforts that would implement the strategic approach described herein. Chapter Five is supported by Appendix B (information necessary to institutionalize the integration of joint data in the personnel systems), Appendix C (draft memos that explain the necessary data requests to the services and external organizations), Appendix D (draft memo of explanation for the individuals that will locally administer the survey to their organization), and Appendix E (protocol versions).

Chapter Six contains conclusions and recommendations.

CHAPTER TWO

Current and Future Joint Missions Require Experienced and Knowledgeable Joint Officers

The goal of a strategic approach to joint officer management is to match the Department of Defense's human resource decisions of the training, education, and experience of officers in joint operations to best support the strategic goals or intent of DoD and the nation. Strategy documents provide the context for joint officer management, as well as offer views of how DoD intends to carry out joint operations. We evaluated legislative guidance on joint officer management contained in Goldwater-Nichols as well as national strategy and DoD military strategy and vision documents to derive stated or implied goals or intent as they relate to joint officer management to achieve success in current and future missions.

Goldwater-Nichols Act

Goldwater-Nichols forged a cultural revolution by improving the way DoD prepares for and executes its mission, and is the driving force behind joint officer management. Title IV of the GNA addresses joint officer personnel policies and provides specific personnel management requirements for the identification, education, training, promotion, and assignment of officers to joint duties. The summarized objectives of the GNA are to enhance joint warfighting capabilities, increase the quality of officers in joint assignments, ensure that officers are not disadvantaged by joint service, ensure that general and flag officers are well rounded in joint matters, enhance the stability and increase the joint experience of officers in joint assignments, and enhance the education of officers in joint matters and strengthen the focus of professional military education in preparing officers for joint duty assignments. The act was a landmark document that changed the way officers are managed, and it provided specific goals that must be met. The GNA has driven changes in the way that officers are educated, trained, and experienced in joint operations, and successes have been achieved.

Since the enactment of the GNA, missions and methods of operation have evolved. The future joint missions will build on the lessons learned from current missions. The strategic guidance in the National Security Strategy, strategic vision documents, and lessons learned point to the imperative to strengthen future joint operations.

National Security Strategy

The President's National Security Strategy establishes strategic goals for the United States. The NSS provides broad strategic guidance on the conduct of joint operations. In it, the President notes, "innovation within the armed forces will rest on experimentation with new approaches to warfare, strengthening joint operations, exploiting U.S. intelligence advantages, and taking full advantage of science and technology."[1] In view of the evolving nature of warfare in the 20th century, the President recognizes the need for future forces to be responsive to different threats. The means that maintaining and achieving strategic goals rests with the ability to change as the threats to the nation have changed. While the NSS does not provide direct inference as to how officers are to be educated, trained, and experienced in joint operations, it does provide broad strategic guidance on how the U.S. military must operate to achieve strategic goals.

---

[1] The White House, *National Security Strategy of the United States of America*, September 2002.

Strengthening joint operations is viewed as an implied strategic intent or goal to support the nation's security.

The NSS states that in order to meet strategic goals, alliances and command structures must be developed, and command structures must be dynamic and adaptable to meet unique mission requirements. Further:

> The alliance must be able to act wherever our interests are threatened, creating coalitions under NATO's own mandate, as well as contributing to mission-based coalitions. To achieve this, we must … streamline and increase the flexibility of command structures to meet new operational demands and the associated requirements of training, integrating and experimenting with new force configurations.…[2]

While the NSS provides overarching strategy, the strategic direction or means to achieve these goals as they relate to joint training, education, and experiencing can be seen in DoD and service documents.

Department of Defense

The Secretary of Defense has provided his vision on the future of joint operations through several documents, including the Quadrennial Defense Review. The QDR provides a vision as to how U.S. military forces will be utilized now and in the future and emphasizes that they must be transformed to meet future challenges. One aspect of the transformation on which change is based is the "pillar of strengthening joint operations through standing joint task force (JTF) headquarters, improved joint command and control, joint training, and an expanded joint forces presence policy…."[3] On joint and combined command and control, the Secretary noted that

---

[2] The White House (2002).

[3] U.S. Department of Defense, *Quadrennial Defense Review Report*, September 30, 2001.

> A joint command and control structure must reside not only at the joint command, but also extend down to the operational service components.... It must be supported by the appropriate doctrine, tactics, techniques, and procedures, as well as a highly trained operational force. Most important, it must develop and foster a joint professional culture, a requirement that presents a significant challenge to service and joint training and professional education programs.[4]

This direction provides a broad-reaching vision on the conduct of joint operations. Further, it implies that the range of personnel who will need to be experienced in joint operations will include operational service component staffs.

Joint Visions 2010 and 2020 are strategic vision documents that seek to provide military strategies for future operations and capture what the U.S. military force must do to realize success in the future joint environment. These Joint Visions provide a common frame of reference as to the future concepts and capabilities of the joint force. Although these are vision documents and may not be strictly viewed as providing strategic goals or intent, they can be viewed as desirable outcomes in support of how the joint force should or are expected to operate in the future. Relating to the increased range of integration of the joint force, Joint Vision 2010 reads:

> A fully joint force requires joint operational concepts, doctrine, tactics, techniques, and procedures—as well as institutional, organizational, intellectual, and system interoperability—so that all US forces and systems operate coherently at the strategic, operational, or tactical levels.[5]

The overall goal of conducting joint operations is to efficiently use the combined capabilities of each service to accomplish the mission. The services do have overlapping capabilities, however, and vision statements add that some are necessary and some redundancies

---

[4] DoD (2001).

[5] Chairman of the Joint Chiefs of Staff, *Joint Vision 2010*, 1995.

will be minimized in the future. To maximize the effectiveness of each service's capability by reducing redundancy requires increased integration of each service's strengths.

The purpose of Joint Vision 2020 is to broadly describe the human aspect required for the joint force to succeed in 2020 and beyond. The vision for the Joint Force in 2020 follows along the same theme as delineated in Joint Vision 2010—i.e., the future force will be more tightly interwoven as a joint force.

> The joint force, because of its flexibility and responsiveness, will remain the key to operational success in the future. The integration of core competencies provided by the individual Services is essential to the joint team. To build the most effective force for 2020, we must be fully joint: intellectually, operationally, organizationally, doctrinally, and technically.[6]

Joint Visions 2010 and 2020 envision the range of joint integration for the force to encompass strategic, operational, and tactical operations.

In recognition of the improved joint operations and the need for continued improvement, the Joint Chiefs of Staff developed a white paper that addresses a path for joint warfare and crisis resolution.[7] Key elements and capabilities from a joint warfighting perspective were identified to support the future of joint warfare. This perspective suggests that, within the operational environment, successful future military operations will continue to require highly qualified personnel, trained to exacting standards and educated to function within a joint force context.[8] To successfully implement joint warfare, an expeditionary and joint team mind-set must be institutionalized. In addition:

---

[6] CJCS (2000).

[7] Joint Chiefs of Staff, "An Evolving Joint Perspective: US Joint Warfare and Crisis Resolution in the 21st Century," white paper, Joint Vision and Transformation Division, January 28, 2003.

[8] Joint Staff, *The Future Joint Force: An Evolving Perspective*, January 28, 2003.

The emerging capabilities required for future joint operations calls for a new culture that emphasizes adaptability in its personnel. To institutionalize this change the Armed Forces of the United States must develop common and comprehensive education, training and exercises across the range of military operations to reinforce the expeditionary and joint team approach to joint warfare.

Further, it is noted that at the operational level, a desired outcome for success is that the joint force in the 21st century will preserve the operational level as the integrating joint force focal point. In addition, the joint force must operate in synchronization with interagency partners at the strategic and operational levels of warfare and crisis resolution. These statements suggest more widely spread joint integration functions to achieve operational objectives.

The services have also advanced their thoughts on the need to operate effectively as a joint team. Service vision documents, which are normally parochial documents for the services, address their respective service's vision for joint operations. The Army's vision addresses where future involvement is likely to be:

The spectrum of likely operations describes a need for land forces in joint, combined, and multinational formations for a variety of missions extending from humanitarian assistance and disaster relief to peacekeeping and peacemaking to major theater wars....[9]

In *Forward...from the Sea*, the Navy addresses joint and combined operations:

No single military service embodies all of the capabilities needed to respond to every situation and threat. Our national strategy calls for the individual services to operate jointly to ensure both that we can operate successfully in all warfare areas and that we can apply our military power across the spectrum of foreseeable

---

[9] U.S. Department of Defense, *The Army Vision: Soldiers on Point for the Nation—Persuasive in Peace, Invincible in War*, Office of the Chief of Staff, 2000.

situations—in peace, crisis, regional conflict, and the subsequent restoration of peace.[10]

The Naval Operating Concept for Joint Operations describes how the Navy will prepare and sustain a capable and ready force, and relates that "the sea base will integrate joint capabilities into a coherent force that will significantly increase the ability of the Joint Force to command and control, project, support, and sustain forces throughout the battle space."[11]

Air Force Vision 2020 addresses its service's mission in joint operations:

> We are partners in our nation's security. We dominate the aerospace domain to facilitate the effectiveness of the Joint Team. Our commitment is firm—to work effectively with soldiers, sailors, marines and coastguardsmen anywhere our nation's interests and its people are at risk. And as members of the Joint Team, our commitment is equally firm to live up to the trust of our multinational partners.

Marine Corps Strategy 21 delineates tenets of success for the future:

> Our goal is to capitalize on innovation, experimentation, and technology to prepare Marine Forces to succeed in the 21st century. Our aims are to evolve maneuver warfare tactics, techniques, and procedures to fully exploit the joint operational concepts articulated in Joint Vision 2020; and evolve our war fighting concepts to enhance our ability to participate as partners in joint and allied concept development and experimentation.[12]

The vision statements of the services support the President's NSS, mirror the vision dictated by DoD, and go beyond the tenets of

---

[10] Department of the Navy, *Forward...from the Sea*, 1994.

[11] Department of the Navy, *Naval Operating Concept for Joint Operations*, undated.

[12] Department of the Navy, *Marine Corp Strategy 21*, 2000.

the GNA in extending the range of training, education, and experience necessary to support effective joint operations.

Finally, recent successes achieved in Operation Iraqi Freedom point to building on the lessons learned in joint operations. In his testimony before the Senate Armed Services Committee on Operation Iraqi Freedom (OIF), the Secretary of Defense indicated that key lessons learned were

> The importance of jointness and ability of U.S. forces to fight, not as individual de-conflicted services, but as a truly joint force—maximizing the power and lethality they bring to bear.[13]

General Tommy Franks, USCENTCOM, stated that the things that worked during OIF included

> The maturing of joint force operations and that Southern/ Northern Watch Operations and Operation Enduring Freedom experiences contributed to the jointness and culture of U.S. Central Command headquarters. There was improved interoperability and C4 and intelligence networking.[14]

Franks added that operational objectives were achieved through integration of ground maneuver, special operations, precision lethal fires, and nonlethal effects. Admiral Edmund Giambastiani's testimony before the House Armed Services Committee on OIF lessons learned included the following comments:

> If you know more and fight together as a joint and combined team, you can act with greater precision, you can rapidly plan

---

[13] Donald H. Rumsfeld, "Lessons Learned" During Operation Enduring Freedom in Afghanistan and Operation Iraqi Freedom, testimony before the Senate Armed Services Committee, July 9, 2003.

[14] General Tommy R. Franks, "Lessons Learned" During Operation Enduring Freedom in Afghanistan and Operation Iraqi Freedom, testimony before the Senate Armed Services Committee, July 9, 2003.

and adapt to fluid situations and you can move about the battle space with far greater effect than what was possible in the past.[15]

Giambastiani also stated that the capabilities that reached new levels of performance that need to be sustained included joint integration and adaptive planning, joint force synergy, and special operations and the integration of special operations forces with convention forces.

Joint Task Forces Are Examples of How Organizational Responses to Missions Affect Officer Qualification

JTFs are the primary organizations for joint operations and far more widely used than they were previously. This organizational structure capitalizes on the capabilities of each service, and the service component commands are frequently used as the basis for them. Other options are standing JTF headquarters or formation of an ad hoc headquarters from various contributors. Many JTFs are designed to accomplish specific objectives and then disestablish. Others are permanent or long standing. The fluidity of these organizations was not a consideration when the GNA legislation was established and the DoD policy for JDAL was set. Frequently, personnel who staff a JTF are in positions on service staffs or on service component staffs and are serving in the JTF as an individual augmentation. Exceptions exist, but in general, personnel in these positions do not receive joint qualifications, either because the position does not qualify or the duration of assignment is limited.

Moreover, beyond the personnel assigned to JTF headquarters, service component commanders and service forces assigned to the JTF also plan and execute joint military operations. As stated earlier, these processes are more closely integrated among the services than such operations were prior to the GNA. All the personnel involved at

---

[15] Admiral Edmund P. Giambastiani, Jr., USJFCOM, Operational Lessons Learned from Operation Iraqi Freedom, testimony before the House Armed Services Committee, October 2, 2003.

this level of activity are in service-specific positions and not eligible for joint qualification, even though the experiences they have may be intensely joint. Further, these positions may require, or benefit from, prior joint experience.

These are the types of issues that are reflective of the current and future joint missions and that arise from mission, organization, and technology changes that a strategic approach is designed to sort out.

Summary

What can be derived from Goldwater-Nichols, the National Security Strategy, the Secretary of Defense's guidance and vision, and lessons learned is that there is a strategic intent that joint operations be strengthened. To achieve this, future operations must interweave more tightly the capabilities of each service, and the need for officers (and enlisted personnel) experienced in joint operations must extend throughout the range of military operations—strategic, operational, and tactical. This was apparent in the preceding excerpts in the assertion of the importance of "U.S. forces to fight...as a truly joint force,"[16] in the emphasis placed on the "jointness and culture" of USCENTCOM,[17] in Admiral Giambastiani's assertion of the importance of officers who know more and thus can fight as a joint team, and in the recognition that a joint professional culture is necessary despite the challenge to service joint training and educational programs. The implication is that as joint operations exist today and continue to evolve, a wider range of officers who are experienced and knowledgeable of joint operations will be needed. Human resource plans and decisions must account for this broader need in support of strengthening joint operations through training, educating, experiencing, and crediting officers for joint operations.

---

[16] Rumsfeld (2003).

[17] Franks (2003).

CHAPTER THREE

Joint Duty Assignment List Management and Joint Officer Development

This chapter provides a descriptive analysis of the service management of the Joint Duty Assignment List. Our analysis suggests differences by service regarding the management of joint assignments for officers. This chapter then describes the management and development of joint officers, as evident from a data analysis of officer files. This description includes actual levels of jointness among officers as well as the management or treatment of officers with different levels of jointness. Quantifying the current availability of jointness is an important aspect of a strategic approach and is a prerequisite to any analytical effort to project the availability of required characteristics in the future. Thus, this chapter is important to the strategic approach because it denotes the current levels of jointness among officers. This chapter concludes with observations based on the data discussed and a consideration of whether Goldwater-Nichols is still necessary.

Service Management of the JDAL

Many officers perceive that the JDAL is constantly increasing and that, while the services have downsized over the past decade, the joint world has continued to expand. Thus, they argue, not only is it difficult to fit a joint assignment into an individual's career path, but it is also difficult to satisfy the joint thirst for their officers when their own service organizations also need people. Figure 3.1 provides the number of billets, by service, on the JDAL over time. These data indicate that the Army, Navy, and Air Force have seen a constant or slightly decreasing number of joint assignments for about the last eight years. The Marine Corps has seen a constant, albeit gradual, increase in its share of the JDAL.[1]

We attribute the common misperception among officers that the JDAL is growing to the likelihood that officers are more aware of additions to the JDAL than they might be to offsetting deletions from the list. Figures 3.2 through 3.5 indicate, by service, the number of billets added to the JDAL, the number of billets deleted from the JDAL, and the cumulative number of existing joint assignments. For example, Figure 3.2 indicates changes to the Army's share of the JDAL. The bars extending upward indicate additions to the JDAL; the bars extending downward denote positions removed from the JDAL. The line indicates the cumulative number of billets representing the Army's share of the JDAL. After the initial years following the GNA, there were some years of net increase and some years of net decrease. However, consistent with the data in Figure 3.1, there has been a general decreasing trend since the mid- to late 1990s. This trend is also apparent in the Navy data (Figure 3.3) and even more apparent for Air Force billets (Figure 3.4). The Marine Corps data,

---

[1] It is true that there has been a relative increase in the percentage of service officers filling external positions. Three of the four services had fewer senior officers (O-4 through O-10) in FY 2003 than they had in FY 1987 as a result of the officer drawdown over those years. As a result, the absolute changes shown in the figure for the number of JDAL positions led to a relative increase in the percentage of officers O-4 through O-10 filling positions on the JDAL. The Army increased from 9 to 12 percent; the Navy from 7 to 9 percent; and the Air Force from 8 to 11 percent. The Marine Corps also saw an increase from 8 to 10 percent as a result of increasing officer positions on the JDAL at a faster rate than it increased O-4 through O-10 officers.

Another point with respect to these absolute numbers should be reiterated. The JDAL by itself does not increase position. Valid positions are moved from internal service organizations to external organizations (e.g., transportation, space) over time or are added to external organizations. These positions may or may not be included on the JDAL as part of a separate process. The services are resourced to fill these external positions as they are for their internal positions. See Harrell et al. (1996).

Figure 3.1
Service Size and Share of the JDAL

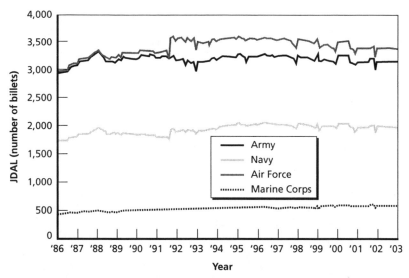

NOTE: Data captured in October of each year.
RAND MG306-3.1

Figure 3.2
Army Starts, Stops, and Cumulative Change to the JDAL

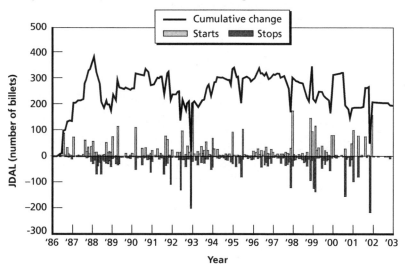

NOTE: Data captured in October of each year.
RAND MG306-3.2

Figure 3.3
Navy Starts, Stops, and Cumulative Change to the JDAL

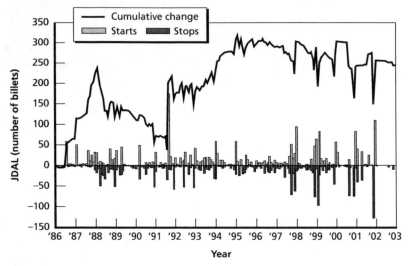

NOTE: Data captured in October of each year.
RAND *MG306-3.3*

Figure 3.4
Air Force Starts, Stops, and Cumulative Change to the JDAL

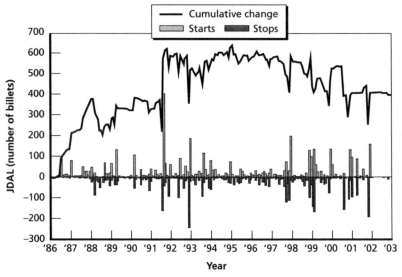

NOTE: Data captured in October of each year.
RAND *MG306-3.4*

Figure 3.5
Marine Corps Starts, Stops, and Cumulative Change to the JDAL

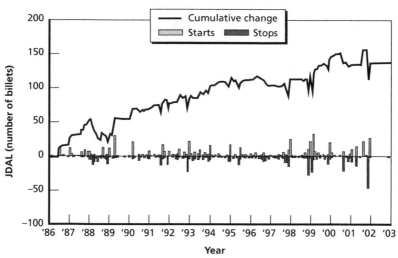

NOTE: Data captured in October of each year.
RAND MG306-3.5

shown in Figure 3.5, indicate a different trend, however. Also consistent with Figure 3.1, the Marine Corps' share of the JDAL is shown here to be steadily increasing.

The Marine Corps representatives were the only service representatives who did not claim difficulty in finding time in an officer's career path for a joint assignment. We see this attitude also reflected in the rate at which services fill joint assignments. Figure 3.6 indicates the rate at which each of the services fills joint assignments. These data indicate a decreasing fill rate for three of the four services, with a cyclical change of fill rates approximating 80 percent of joint assignments. The Marine Corps, however, exhibits a markedly different pattern: The Corps is increasing the rate at which it fills joint assignments, to between 90 and 95 percent. Taken in concert with the data from the prior figures, we see a distinct difference between the Marine Corps and the other services. The Marine Corps is not only increasing its share of assignments on the JDAL but is also filling its

Figure 3.6
JDAL Fill Rates, by Service

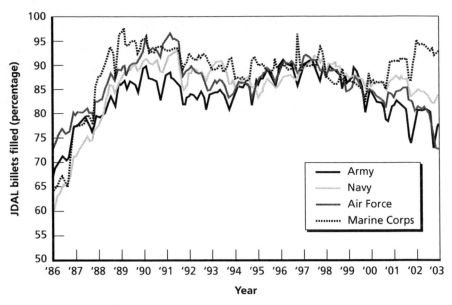

NOTE: Data captured in October of each year.
RAND *MG306-3.6*

joint assignments at a higher rate, whereas the other services are both decreasing their joint assignments and filling existing joint assignments at a considerably lower rate.

Also of interest is the degree to which the services use joint assignments efficiently in the course of officer development. Regardless of the value placed on the experience gained in a joint assignment, the current system grants an officer credit for joint duty only if he remains in a JDA for a sufficient period.[2] Figure 3.7 indicates that approximately 80 percent of Marine Corps officers serving in a JDA remain in their position sufficiently long to receive joint duty credit.

---

[2] COS officers and general and flag officers must serve two years in an assignment to receive credit. Other officers must serve three years to receive credit. Cumulative credit from multiple assignments can also provide credit to officers. This is also discussed in Appendix A.

Figure 3.7
Percentage of JDAL Officers Receiving Joint Credit, 1987–2001

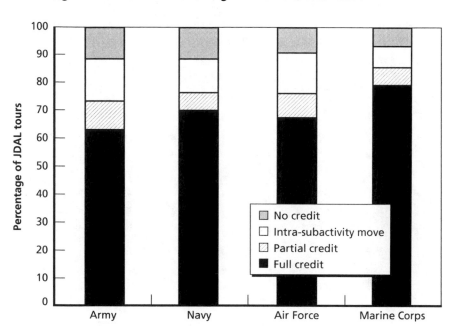

RAND *MG306-3.7*

In contrast, fewer than two-thirds of Army officers receive full joint credit for their time in a JDA. The Air Force and the Navy are more efficient than the Army but are still notably less efficient than the Marine Corps.

In summary, we understand from service personnel responsible for officer development that, from the service perspective, it is generally very difficult to provide officers with a joint assignment. Other than the need to "check the box" as having had such an assignment, there is little perceived benefit to the officer or to the service. The Marine Corps does not claim such difficulty in providing its officers to joint organizations. The data indicate that past management of officers and joint duty assignments generally support these perspectives: The Marine Corps is the only service that is increasing its share of the JDAL, increasing the rate at which it fills joint assignments, and attempting to maximize the number of officers that receive credit for their joint assignments. The other services are decreasing their joint share, decreasing the rate at which they fill joint assignments, and providing joint credit to smaller proportions of officers serving in joint assignments.

Different Views of Filling the JDAL

In this last look at the current JDAL, we examine how positions have been filled by officers with different backgrounds or experiences. This is intended to portray the types of officers who are filling JDAL positions, at a macro level. To do this, we extract the assignment history for each current position on the JDAL or determine what happens to officers who have served in the position. For example, greater than 20 percent means that every time the position was filled (one or more), an officer with the characteristic shown filled the position. As shown in Figure 3.8, 60 percent of positions are filled by critical occupation specialty (COS) officers at least 20 percent of the time and more than 30 percent of positions are filled by COS officers 100 percent of the time.[3] In other words, approximately one-third of positions are filled consistently with a combat arms or unrestricted line officer.[4] When we consider the positions that are filled with officers who have received joint education or previous joint experience, we see that only 10 percent of positions have JPME II graduates 100 percent of the time and only 6 percent of positions have someone with a previous joint assignment 100 percent of the time. When we consider the subsequent assignments of officers, we see that there are very few positions from which officers retire 100 percent of the time they are filled. In other words, most officers who serve in joint assignments are

---

[3] COS includes combat arms occupations from each of the services. This is discussed further in Appendix A.

[4] See later section on management frameworks. These positions are likely to be used in a leadership succession model and thus are more likely to have more rapid turnover, even though the GNA limits the number of COS officers who can serve just two years and still receive joint credit to 12.5 percent of serving JSOs and JSO nominees.

Figure 3.8
Fill of the JDAL, by Officer Characteristic

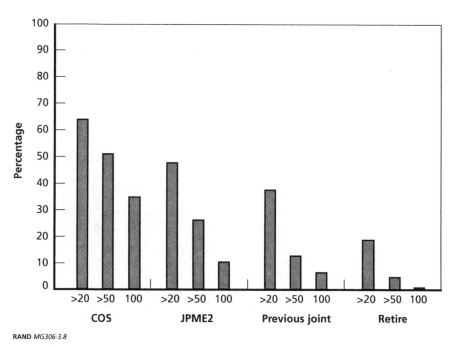

taking their experience from that position to subsequent service or joint assignments; only a few positions are consistently filled by officers who will retire from that position.

Joint Officer Development: The Service Perspective

Service personnel responsible for officer development generally express difficulty in exposing their officers to joint education and joint assignments. Early acculturation of officers is focused on "greening" or "bluing," with some exposure in early officer education to the concept of jointness. Three of the four services (all but the Marine Corps) claim that it is difficult to divert an officer from service assignments to a joint assignment to gain experience beyond the initial exposure. Personnelists cite the important service assignments that an officer needs to complete for his development in order to maximize his likelihood of receiving O-5 or O-6 command. After a command assignment, the officer is perceived to have valuable experience that is needed by service organizations. In addition to the concerns that these personnel expressed, the semantics used while describing joint assignments confirm these attitudes: Officers are "diverted" or "ripped" from service assignments in order to complete a joint assignment. Among those personnel who are responsible for officer development, while service assignments are valuable developmental opportunities that contribute to a better-developed officer who is thus more valuable to the service, joint assignments are "wedged into" a career path so that an officers can "check the box" for having received an experience that is generally not perceived to benefit other service assignments.

For those officers who are sent to joint assignments, Army and Marine Corps selection of officers for joint assignments appears to be more proactive and planned than in the case of the Navy. Many of the Air Force officers that we interviewed in the context of this research perceive the benefit of joint experience to be so considerable that joint assignments are considered highly attractive and available only to select officers or to those who enjoy the benefits of a strongly influential mentor. In contrast, Navy officers tend to minimize the importance of a joint assignment, reflecting a service perception that joint experience does not improve the performance of officers in Navy assignments. Further, our interviews suggest that Navy officers may tend to believe that they can learn jointness without formal exposure to the joint environment. Officers from the other services appear more likely to acknowledge the value of the joint experience in their performance of other assignments, but differ in their perception of the likelihood that officers will gain such experience during the course of a career.[5]

---

[5] In a later section of this report, we discuss how management frameworks may have reinforced such attitudes and how these frameworks might be used to analyze and assess changed practices.

Regardless of some positive perceptions, and despite the fact that GNA mandates jointness for career success at higher ranks, we did not see a general cultural acceptance of jointness or a unanimous acceptance of jointness as a key to success. Instead, we heard officers state that they did not join the Marine Corps to be joint, and that "staying blue" ("haze gray and under way") reflects Navy attitudes. The Army and Air Force focus on internal assignments supplemented with a joint "ticket punch." The sought-after field grade assignments are service commands. These attitudes voiced during our interviews were also confirmed by recent comments reflecting past views from the Army's Vice Chief of Staff: "In the Army I grew up in, we largely thought of ourselves as the supported service—we thought in terms of what the other services could do for us. Most of us paid lip service to jointness—to joint training, joint assignments, joint education— because success was judged in Army terms."[6]

However, we do not want to paint an overly bleak picture. There are examples of communities that are more joint in outlook and management than the mainstream communities. For example, special operations communities appear to have more in common with the special operations communities of other services than with other communities in their service. Similar outlooks also appear to exist for other specialized communities such as intelligence and communications. And as the subsequent section demonstrates, even for the mainstream communities, an increasing number of officers are joint. Perhaps the best summary of the state of joint officer training and development is to examine outcomes. How joint is the officer corps?

Joint Officer Management: Actual Outcomes

Since 1986, officers have been assigned to positions on the JDAL, have attended JPME II, have had other assignments and educational opportunities, and have been promoted. Moreover, many of these

---

[6] GEN George W. Casey, Jr., Army Vice Chief of Staff, speech to the West Point Society and the National Capital Winter Luncheon, January 21, 2004.

officers have confronted the decision to stay or leave in the interven-
ing years. Descriptive data about such officers are presented each year
as part of the Secretary of Defense's Annual Report to the President
and the Congress, but those data are a snapshot at a particular point
in time.

The data we present below are longitudinal data. In other
words, we keep track of what happened to officers who have ever
served in a joint assignment or gone to a joint school so that we may
see how experience and education accumulate in the officer corps and
examine what their career outcomes were.[7] What are the differences
between officers who have ever been "joint" and those who have not?

Four general observations emerge from our look at this data:

- In general, the data give evidence that jointness is accreting in
  the officer corps. Officers as a group are more "joint" with each
  passing year. Officers who are JSOs, who are joint qualified, or
  who have some joint experience advance in grade and stay in
  service at rates sufficient to have increased overall joint content
  of the officer corps over time. This is more accurately stated for
  certain grades, occupations, and services then for others.
- One can draw different inferences by looking at the data in different ways. For example, if one examines the entire officer corps
  for grades O-4 and above, the fact that the corps has accumulated about 15 percent of officers with at least some joint experience does not seem high.[8] However, the denominator in the

---

[7] We created this database to support our modeling approach that is based on stocks and
flows of officers over time. We used the Joint Duty Assignment Management Information
System (JDAMIS) files, including the historical files, and merged them with a longitudinal
database of officers maintained by RAND. Thus we could link people who had ever served
in a JDAL position with their career outcomes. We count only officers who have served in
positions on the JDAL for some or all of a creditable tour. In terms of joint experience, as
opposed to joint qualification, we are undercounting because the JDAL does not include all
positions external to a service, positions in service components that provide joint experience,
temporary duty assignments to JTFs, and other potential positions.

[8] We parsed joint experience into three groups. The first are those who have received full
credit for a joint tour but are not JSOs. The second group are those who are JSOs. The third
group are those who at a point in time are not in the first two groups but have some joint

equation consists of many officers in occupations not inherently joint (e.g., health) and is more heavily weighted toward O-4s who have had less time to be joint. As one focuses on certain occupations (e.g., intelligence, tactical operations) or on particular grades at which a higher cumulative opportunity to become joint has existed (e.g., O-6), one sees a far rosier picture of the accrual of jointness in the officer corps.

- Increases in jointness have consistently occurred each year. However, a cursory look at the data indicates that such increases are leveling off, i.e., becoming asymptotic at current levels. How joint the officer corps can be is dependent on the opportunity to have a joint duty assignment and to attend JPME II. The seats for the latter are limited, and the number of the former is also limited. It may be that given these constraints, underlying job and educational durations, and continuation and promotion rates, the ability to increase jointness further in the officer corps in the future may not exist absent changes in the number and duration of school and assignment seats. These limits can be explored with career models of the type we discuss later in this report.

- While there are observable differences in behaviors and outcomes between those with and without joint experience, these differences are more apparent when considering leaving data than advancement data,[9] and many of these differences either may not be significant or may result from other factors than jointness. For example, promoting more to O-4 in one year than in other years can change the joint content at that grade for one or more years.

---

experience in a qualifying position. For example, they may be currently serving in such a position or they may have received partial credit for past service. We also track officers who have no joint experience in a qualifying JDAL position. We will differentiate among these groups for some of the figures and analysis in this section.

[9] We consider officers who leave to be those that exist in the data set in a given year but are no longer officers in the subsequent years. We do not attribute a reason for their leaving. Advancement represents those officers who exist at pay grade x in one year and exist in the data set at pay grade x+1 in the subsequent year.

We present some of these data below. Generally, we describe what we are observing and offer an interpretation of what is being observed. We first look at the experience and education data and then review the longitudinal data dealing with advancement and turnover.

The Accumulation of Joint Experience Among Officers

The percentage of officers (O-4 and above) who have had at least some previous joint experience has increased over time. Excluding health occupations, as of 2002 between 30 and 40 percent of officers in each service have had at some time been assigned to a JDA as shown in Figure 3.9. We include in these data officers who have par-

Figure 3.9
Non–Health Care Officers, by Service, with a Joint Assignment
(pay grades O-4 and above)

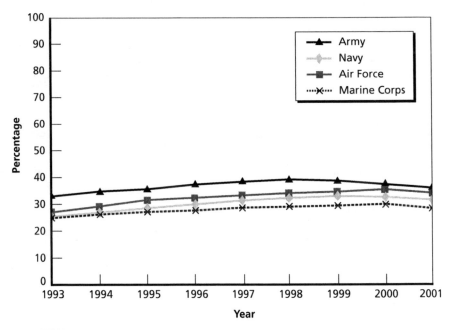

tial credit or who are or were "currently serving" at the time represented by the data.[10] As we examine below, grade and occupational composition affect this overall data. In particular, including O-4 biases joint content downward and largely accounts for decreases in 2000 and 2001.

Figures 3.10 through 3.13 present more detailed pictures of each of the services and indicate that, when the percentage of officers with joint experience is considered by pay grade, certain grades appear increasingly joint. While the portion of O-4s with a joint experience remains both relatively low and relatively stable,[11] generally the percentages of O-5s and O-6s with joint experience have increased since the early 1990s, and in the case of O-6s, continue to increase.

Not surprisingly, the percentage of O-4s who have ever served in a JDA holds reasonably steady between 15 and 25 percent in each service. There is less opportunity to serve at this grade and to accumulate service. The data for O-5 and O-6 are largely about accruing joint service in prior grades. However, the accumulation may have reached limits because of the constraints on the numbers of JDAL positions and JPME II school seats. Our analysis indicates that the O-6 data in particular represent both an increase in O-6s serving in a first JDA as well as the cumulative effect of earlier increasing service

---

[10] Officers who serve in a JDA but do not stay sufficiently long (two years for COS officers and general and flag officer, three years for all other officers) receive partial credit, which is cumulative. Thus, officers who serve in multiple JDAs with insufficient tenure each time can receive joint credit for a cumulative number of total days served. This is also discussed in Appendix A.

[11] In some services, the percentage of O-4s with at least some joint experience is decreasing. Frequently, this is because the numerator and denominator are changing (more or fewer O-4s in a given year) at different rates. Frequently, we are observing cohort dynamics that linger from the officer entry patterns prior to the drawdown, the drawdown entry and continuation patterns, and the post-drawdown continuation and promotion patterns. Because of these cohort patterns, we have previously recommended that performance data such as those for year-to-year promotions should be smoothed, for example, with a moving average. See Harrell et al. (1996).

Figure 3.10
Non–Health Care Army Officers with a Joint Assignment

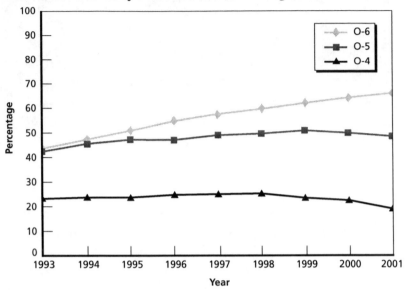

Figure 3.11
Non–Health Care Navy Officers with a Joint Assignment

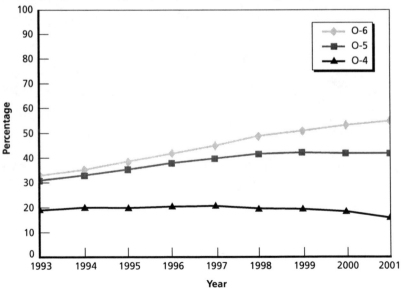

Figure 3.12
Non–Health Care Air Force Officers with a Joint Assignment

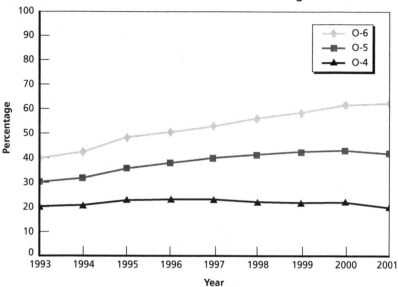

Figure 3.13
Non–Health Care Marine Corps Officers with a Joint Assignment

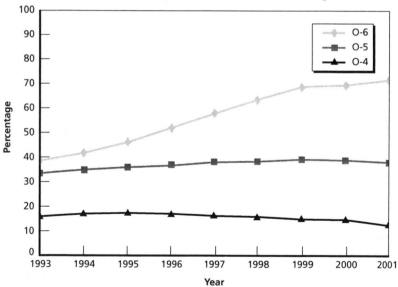

at grade O-5. This pattern is especially evident among non–health care O-6s for each of the services. In Figures 3.14 through 3.17, we portray O-6 data for the four services. For each, the data are arrayed by those who have had a qualifying JDAL assignment, for those who are JSO, and for those who have partial credit or who are currently serving.

From this we see that Navy O-6s are the least likely to have had joint assignments, but the percentage of Navy O-6 officers with a joint experience still increased from 30 percent in 1993 to over 50 percent in 2001. The Marine Corps has seen the most dramatic increase, from slightly more than a third of their O-6s having had joint experience as of 1993 to almost three-quarters in 2001. One possible reason for this, as seen in the data in the earlier section, is that the Marine Corps over this entire period had been adding to its portion of the JDAL, filling JDAL positions at a higher rate, and achieving joint tour completion at a higher rate than the other services.[12]

Joint Experience Within Occupations[13]

Looking at the O-5 and O-6 combined officer population by occupation yields further understanding of which officers tend to have had joint experience. We include all three groups identified previously in

---

[12] Nominally, the Marine Corps also has a higher proportion of officers in tactical operations who turn over more rapidly in joint assignments as COS officers. As of October 2004, about 46 percent of USMC officers were in tactical operations compared with 32 to 38 percent of officers in the other services. The Marine Corps has no officers in the health care occupation that accounts for 16 to 20 percent of officers in the other services. With the health care occupation removed, as was done with these data, the proportions of tactical operations officers are comparable in all services.

[13] We used the standard one-digit officer occupations from the DoD Occupational Conversion Index (DoD 1312.1-I), which groups military service occupations in a logical and consistent structure. We excluded health care from our summary analysis and show data for the five largest (of nine) occupational groups. The average data exclude only health care.

Figure 3.14
Army O-6 Non–Health Care Officers

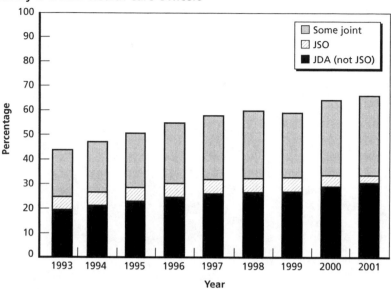

Figure 3.15
Navy O-6 Non–Health Care Officers

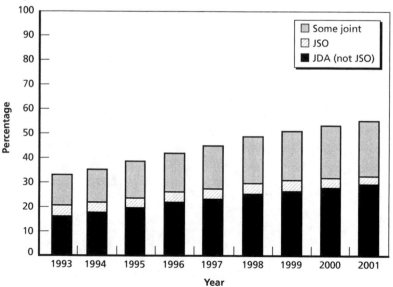

Figure 3.16
Air Force O-6 Non–Health Care Officers

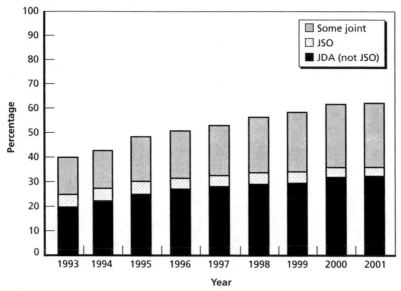

Figure 3.17
Marine Corps O-6 Non–Health Care Officers

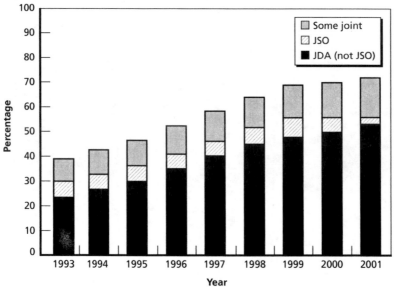

these data. Figures 3.18 through 3.21 indicate that, for each service, intelligence officers are considerably more likely to have experienced a joint assignment. Tactical operations officers are especially of interest, since they are the warfighters; many would maintain that these officers are the most likely to benefit from an understanding of jointness. Additionally, tactical operations officers are the future leaders, those most likely to be promoted to be general or flag officer, and thus those most likely to be required to have had joint experience.[14] These figures indicate that, while tactical operations officers are considerably less likely than intelligence officers to have been assigned to a joint billet, they are increasingly likely to have had such an assignment. That is, the data generally indicate an overall positive trend, with 50 to 60 percent of O-5 and O-6 tactical operations officers having had at least some joint experience, as of 2001. However, in recent years the numbers have stabilized.

Tactical Operations Officers with Joint Experience

Given that tactical operations officers are the majority of future military leaders, it is worthwhile to confirm the accumulation of joint experience among these officers.[15] Figure 3.22 confirms that the amount of joint experience among these officers has increased but has begun to stabilize, especially among Air Force and Army officers. The amount of joint experience among tactical operations officers is approximately 10 percent higher than that of the larger population of non–health care officers, displayed earlier in Figure 3.9.

---

[14] Of the approximately 900 active component general and flag officers, about 660 are in the line communities (tactical operations) that will lead and command at the most senior levels. Those in professional, technical, and support communities serve in important positions and functions but are not likely to rise to the highest levels of DoD as part of their career paths. See Margaret C. Harrell, Harry J. Thie, Peter Schirmer, and Kevin Brancato, *Aligning the Stars, Improvements to General and Flag Officer Management,* Santa Monica, Calif.: RAND Corporation, MR-1712-OSD, 2004.

[15] See, for example, Harrell, Thie, et al. (2004). In that analysis, we observed that approximately two-thirds of general and flag officers were tactical operations officers.

**Figure 3.18
Army O-5 and O-6 Officers with Joint Assignment,
by Occupation**

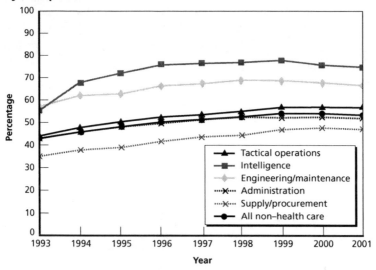

**Figure 3.19
Navy O-5 and O-6 Officers with Joint Assignment,
by Occupation**

Figure 3.20
Air Force O-5 and O-6 Officers with Joint Assignment,
by Occupation

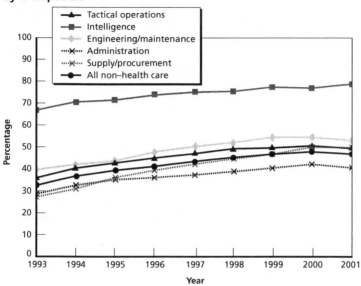

Figure 3.21
Marine Corps O-5 and O-6 Officers with Joint Assignment,
by Occupation

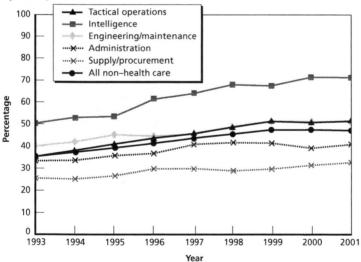

Figure 3.22 indicates the relative increase in joint experience among tactical operations officers, but it is important to note that most of the increase has occurred among O-6 tactical operations officers, as indicated in Figures 3.23 through 3.26.

Figure 3.22
Tactical Operations Officers with a Joint Assignment, by Service (pay grades O-4 and above)

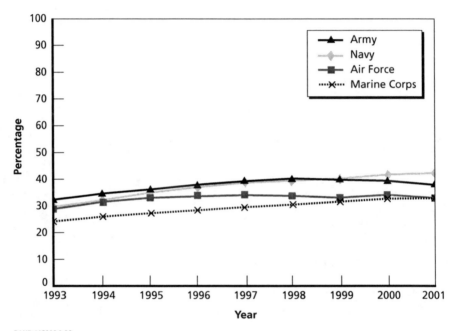

RAND *MG306-3.22*

Figure 3.23
Army Tactical Operations Officers with a Joint Assignment

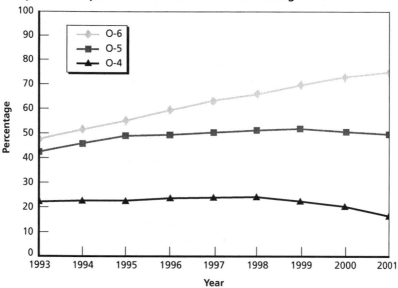

Figure 3.24
Navy Tactical Operations Officers with a Joint Assignment

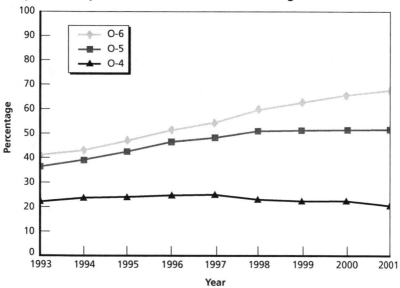

Figure 3.25
Air Force Tactical Operations Officers with a Joint Assignment

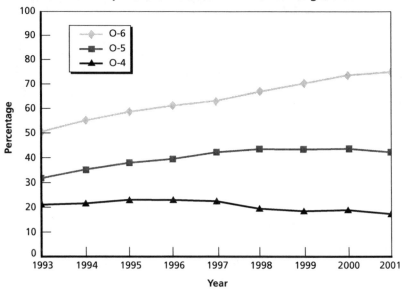

Figure 3.26
Marine Corps Tactical Operations Officers with a Joint Assignment

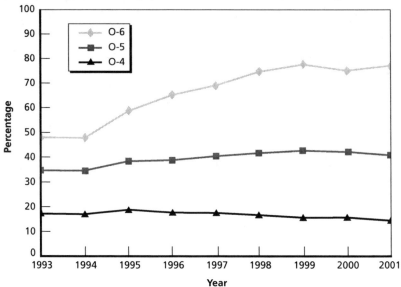

Figures 3.23 through 3.26 indicate that the share of tactical operations O-4 officers with joint experience generally has declined slightly and that the percentage of tactical operations O-5 officers has increased but stabilized; the share of tactical operations O-6 officers continues to increase. Approximately 75 percent of tactical operations O-6s (65 percent for the Navy) had experienced a joint assignment in 2001. However, in three of the four services, the portion of O-6s who have been exposed to a joint assignment but received only partial credit is likely increasing. Figures 3.27 through 3.30 further describe the tactical operations officers with joint experience. For the first three services, these charts indicate a decreasing portion of JSOs, a slightly increasing portion of officers who have received joint credit (without being a JSO), and an increasing portion of officers who have received only partial joint credit or who were "currently serving" at the time of the data point. Had a significant portion of the "currently serving" officers continued in assignment long enough to receive credit, however, the JDA portion of the bars would likely indicate a larger annual increase. These figures suggest that, for three of the services, while 65 to 75 percent of the tactical operations O-6s have been exposed to joint experience, only about 40 percent of them been accredited with jointness (including JSOs).

For the Marine Corps, the portion of tactical operations officers who have received joint credit has steadily increased, while the portion of officers currently serving or with partial credit has remained consistent, suggesting that many of those officers do subsequently receive full joint credit. This is consistent with the previously discussed Marine Corps effort to increase its share of the JDAL, increase the rate at which it fills those positions, and to provide its officers with joint credit.

Figure 3.27
Army O-6 Tactical Operations Officers

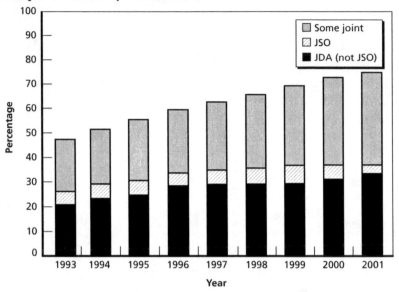

Figure 3.28
Navy O-6 Tactical Operations Officers

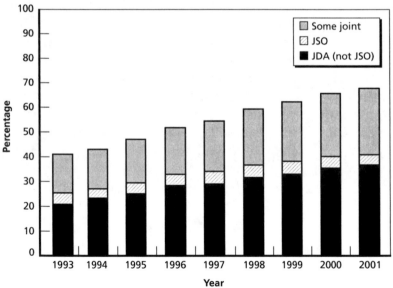

Figure 3.29
Air Force O-6 Tactical Operations Officers

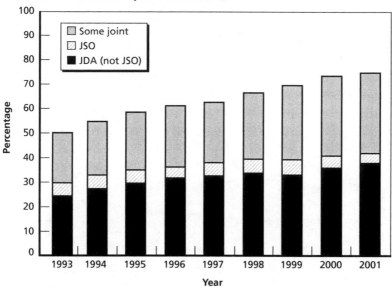

Figure 3.30
Marine Corps O-6 Tactical Operations Officers

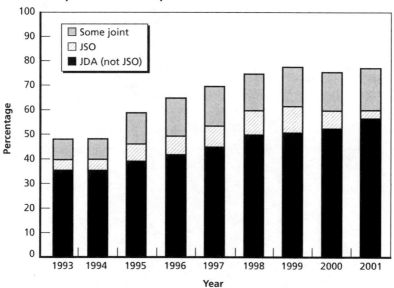

Officers with Joint Education (JPME II)[16]

The data suggest that the services have been exposing an increasing number of O-5 and O-6 non–health care officers to joint education at some point in their career. Even so, Figure 3.31 indicates that, as of 2001, fewer than 30 percent of officers in grades O-5 and O-6 had received joint education, up from fewer than 20 percent in 1993. As pointed out earlier, among other factors that limit JPME II attendance are the number of available seats and their throughput that is affected by course duration, particularly at JFSC.[17]

Figure 3.31
O-5 and O-6 Non–Health Care Officers with JPME II

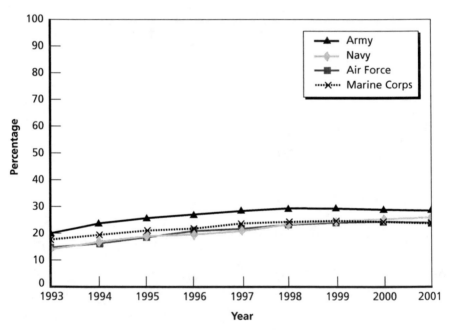

RAND *MG306-3.31*

---

[16] Appendix A includes a brief description of joint education.

[17] The 2005 NDAA decreased the duration of the JPME II course of instruction at JFSC from 12 weeks to 10 weeks. JFSC officials have indicated that they will begin to conduct four sessions per year, with the maximum capacity of 255 students per session, an increase of 120 seats annually.

The services send officers from several occupational groups to joint education (Figures 3.32 through 3.35). This may reflect an attempt to satisfy the GNA requirement that half of the officers in joint assignments complete joint education, as well as the need to produce JSO. Should the requirement be changed (as proposed) such that promotion to O-7 requires JPME II (the most common way for satisfying a prerequisite for achieving JSO status), completion of JPME II becomes more important. The following figures indicate the proportion of officers from different occupations who have completed JPME II. All non–health care officers are also charted, as a comparison basis; this line is inclusive of the other categories shown.

Figure 3.32
Army O-5 and O-6 Officers with JPME II, by Occupation

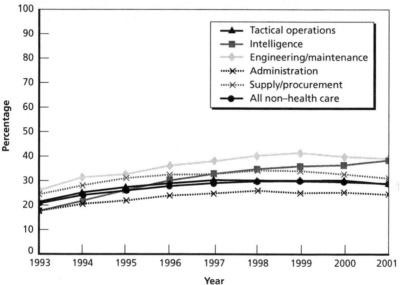

**Figure 3.33
Navy O-5 and O-6 Officers with JPME II, by Occupation**

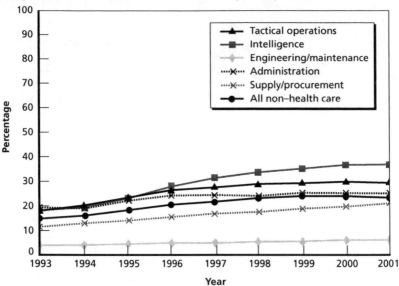

**Figure 3.34
Air Force O-5 and O-6 Officers with JPME II, by Occupation**

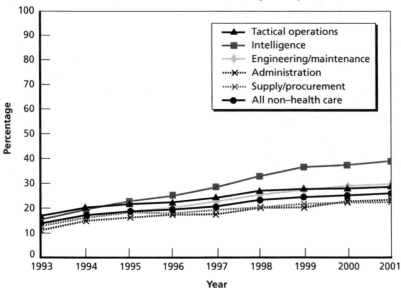

Figure 3.35
Marine Corps O-5 and O-6 Officers with JPME II, by Occupation

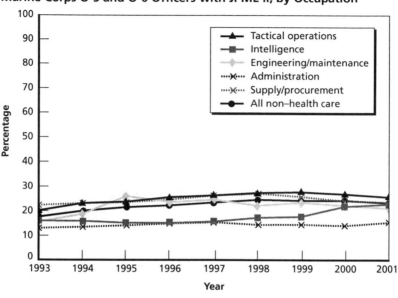

RAND *MG306-3.35*

Figures 3.32 through 3.35 demonstrate empirically that, with the exception of Marine Corps, tactical operations officers are generally less likely than intelligence officers and others to have attended JPME II. Figures 3.36 and 3.37 provide the percentage of tactical operations O-5s and O-6s who have attended JPME II. The tactical operations O-5s appear to have attended JPME II at about the same rate as O-5s and O-6s overall in the services (Figure 3.31); the portion of such officers has stabilized; and there are few differences among the services. However, the percentage of tactical operations O-6s with JPME II is increasing for each of the services. Approximately 35 to 45 percent of tactical operations O-6s in each of the services have attended JPME II, with the Army and the Air Force having the highest proportion of graduates in 2001.

Figure 3.36
O-5 Tactical Operations Officers with JPME II

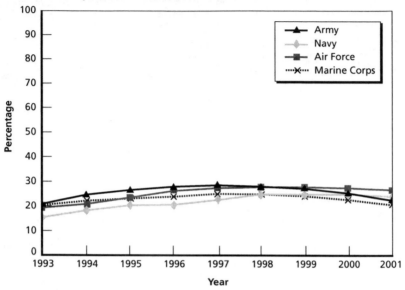

Figure 3.37
O-6 Tactical Operations Officers with JPME II

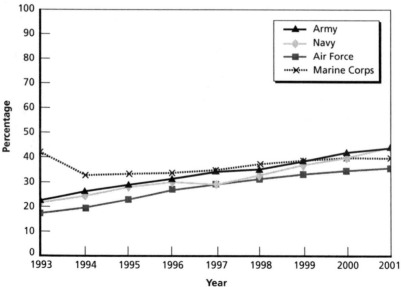

Behavior of and Outcomes for Officers with Joint Experience: Leaving Rates of Officers

The preceding discussion has described the evolving representation of officers with joint experience and joint education, by service, by pay grade, and by occupation. The data previously displayed indicate an increasing number of officers, especially O-6s, who have been assigned to a joint assignment, although fewer of these same officers have received joint education. These increasing trends do not indicate more officers overall being assigned to joint assignments; the number of joint assignments has stayed relatively steady, and the rate at which the services fill the joint assignments has decreased for all services except the Marine Corps. Instead, the extent to which jointness is reflected among officers is a reflection of changes in the way officers (i.e., those with a joint assignment) have been managed, assigned, and promoted and how officers themselves behave with respect to their careers.

What, then, has happened to the officers who did serve in joint assignments? The following discussion explores the turnover and advancement rates for officers with differing levels of joint duty experience, compared with their non-joint peers. Figures 3.38 through 3.45 display the turnover of non–health care officers, based on their pay grade and whether they had joint experience of a particular type. We call this the leaving rate and calculate it by dividing the number of officers with a particular characteristic who left in the year indicated by the number with the same characteristic at the end of the previous year.

These figures indicate that the four services have experienced similar turnover patterns of their personnel. In general, O-5 and O-6 officers without joint experience are more likely to leave than officers who served in a JDA.

Figure 3.38
Leaving Rate of Army O-6 Officers, by Joint Experience

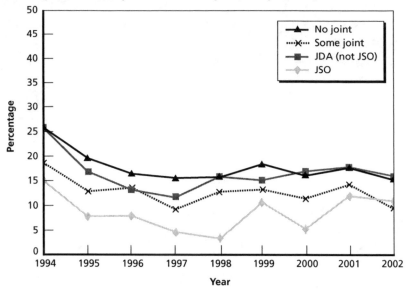

Figure 3.39
Leaving Rate of Army O-5 Officers, by Joint Experience

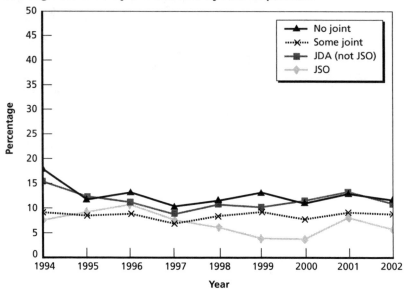

Figure 3.40
Leaving Rate of Navy O-6 Officers, by Joint Experience

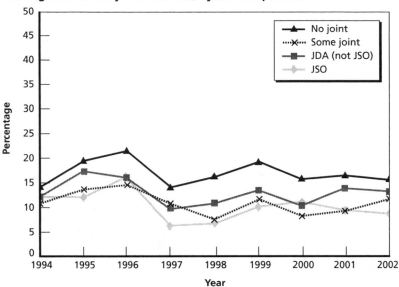

Figure 3.41
Leaving Rate of Navy O-5 Officers, by Joint Experience

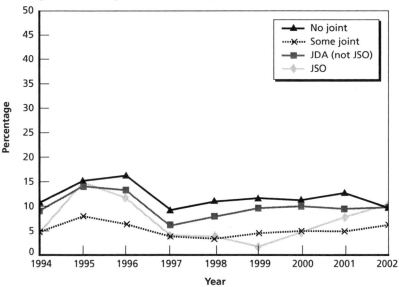

Figure 3.42
Leaving Rate of Air Force O-6 Officers, by Joint Experience

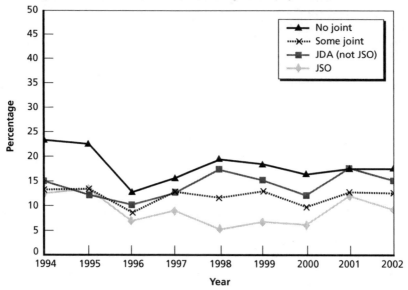

Figure 3.43
Leaving Rate of Air Force O-5 Officers, by Joint Experience

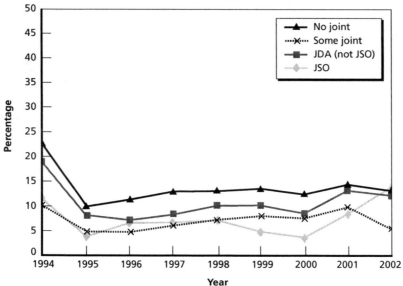

Figure 3.44
Leaving Rate of Marine Corps O-6 Officers, by Joint Experience

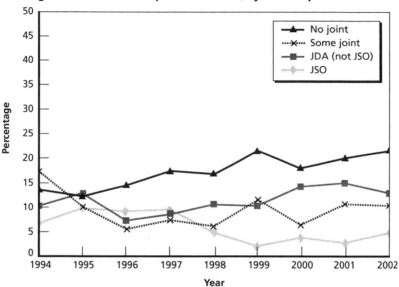

Figure 3.45
Leaving Rate of Marine Corps O-5 Officers, by Joint Experience

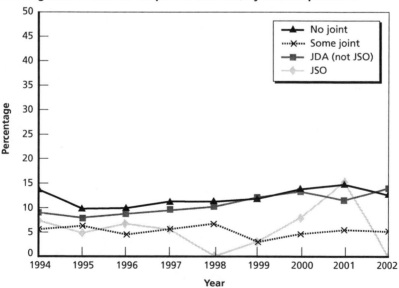

Figure 3.38 through 3.45 indicate that officers with joint experience are not necessarily leaving the services at higher rates. Indeed, officers who lack any joint experience appear to leave the service at rates at least similar to (and generally higher than) officers who have full joint credit, who are JSOs, or who have some joint experience. This outcome varies somewhat when we consider the services' line officers, or warfighters. Figures 3.46 through 3.53 indicate that while in most cases officers without joint experience leave at higher rates, there is not a distinguishable difference among Marine Corps O-5 tactical operations officers. In most cases, tactical operations officers who are JSOs appear to be retained at relatively higher rates. However, many of the differences between the different groups of tactical operations officers are relatively small.

**Figure 3.46
Leaving Rate of Army O-6 Tactical Operations Officers,
by Joint Experience**

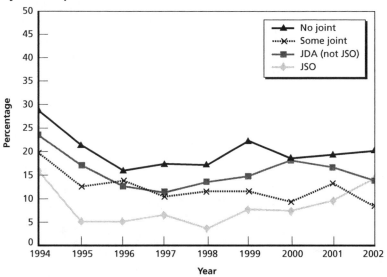

**Figure 3.47
Leaving Rate of Army O-5 Tactical Operations Officers,
by Joint Experience**

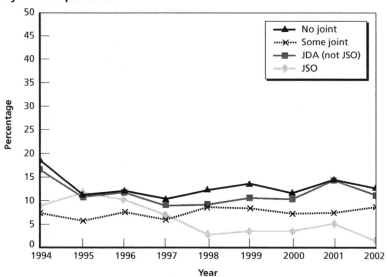

**Figure 3.48
Leaving Rate of Navy O-6 Tactical Operations Officers,
by Joint Experience**

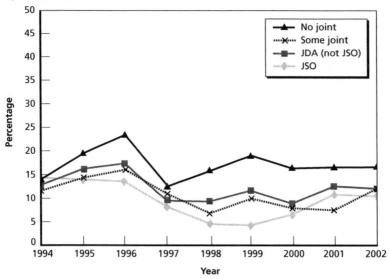

**Figure 3.49
Leaving Rate of Navy O-5 Tactical Operations Officers,
by Joint Experience**

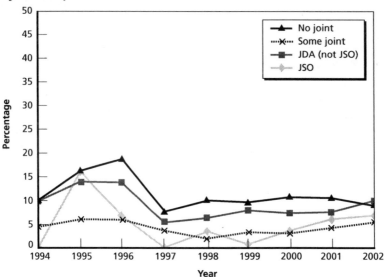

Figure 3.50
Leaving Rate of Air Force O-6 Tactical Operations Officers,
by Joint Experience

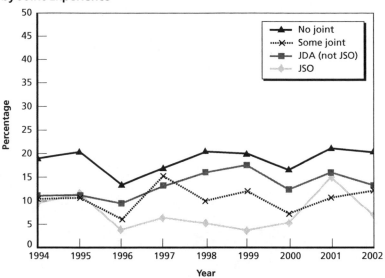

Figure 3.51
Leaving Rate of Air Force O-5 Tactical Operations Officers,
by Joint Experience

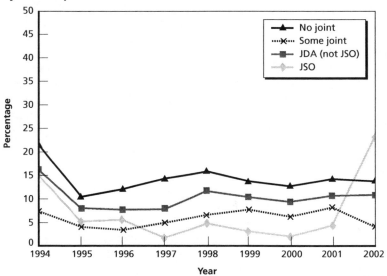

**Figure 3.52
Leaving Rate of Marine Corps O-6 Tactical Operations Officers,
by Joint Experience**

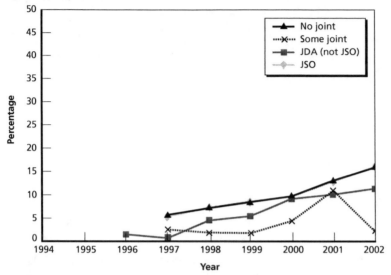

NOTE: The 1994–1996 data of Marine Corps O-6s are incomplete.
RAND *MG306-3.52*

**Figure 3.53
Leaving Rate of Marine Corps O-5 Tactical Operations Officers,
by Joint Experience**

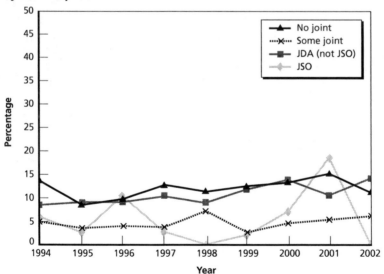

RAND *MG306-3.53*

Behavior of and Outcomes for Officers with Joint Experience: Advancement Rates of Officers

It is also important to consider whether those officers with joint service who remain in the military are being advanced at a rate consistent with their non-joint experienced peers. Figures 3.54 through 3.61 explore advancement rates to the next grade for non–health care officers with a prior joint assignment and those lacking joint experience, for each service. The data are calculated similarly to the turnover rates. The numerator is the number of officers with a characteristic who were promoted to that pay grade in the year shown. The denominator is all officers (not just promotion eligible officers) with the characteristic serving in the lower pay grade in the prior year. We are calculating the flow of officers who advance to a higher grade and contribute to the new inventory of officers in that grade in the next year. This is very different from how the Office of the Secretary of Defense and the services calculate promotion opportunity, which is why we called it an "advancement" rate.[19]

No clear story emerges from these figures. At some times, those with joint experience of a certain type will advance at a higher rate than those with other types of joint experience.

---

[19] Promotion opportunity calculations are based on the number of individuals being considered for promotion—those "in zone"—whereas our advancement rate is based on the total number of people in the pay grade.

Figure 3.54
Advancement of Army O-6 Officers, by Joint Experience

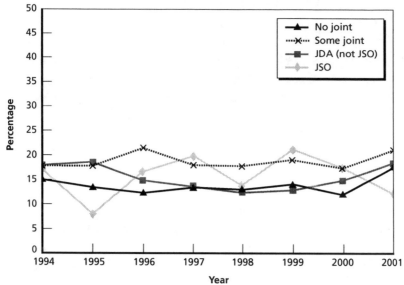

Figure 3.55
Advancement of Army O-5 Officers, by Joint Experience

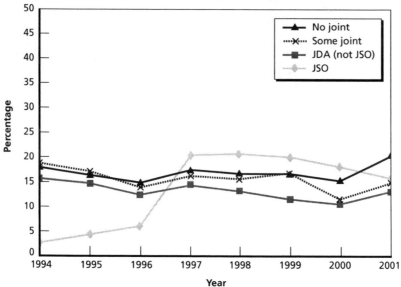

Figure 3.56
Advancement of Navy O-6 Officers, by Joint Experience

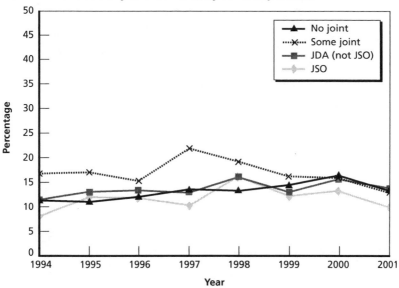

Figure 3.57
Advancement of Navy O-5 Officers, by Joint Experience

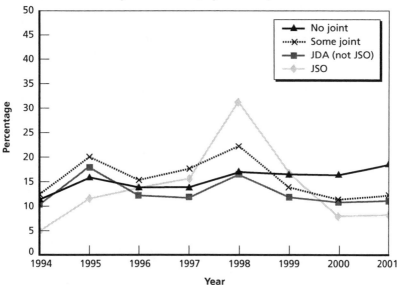

Figure 3.58
Advancement of Air Force O-6 Officers, by Joint Experience

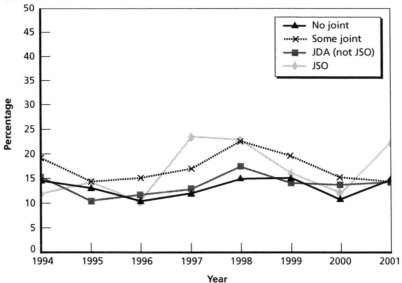

Figure 3.59
Advancement of Air Force O-5 Officers, by Joint Experience

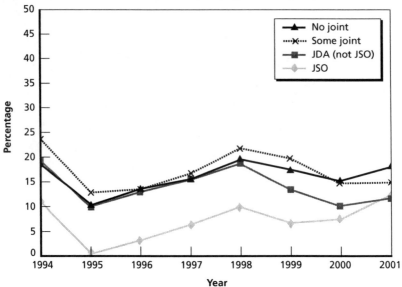

Figure 3.60
Advancement of Marine Corps O-6 Officers, by Joint Experience[20]

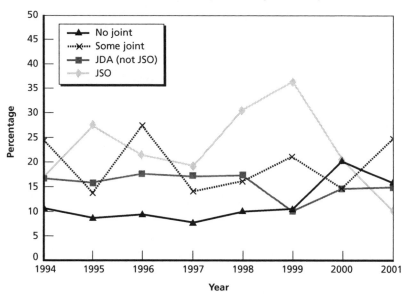

RAND *MG306-3.60*

---

[20] As pointed out in an earlier footnote, these data are hard to interpret, and as stated previously, there is no clear story. In particular, annual promotion data in Figure 3.60 is not independent of prior years or of separation patterns. For example, in that figure, higher than average numbers of JSO were promoted in 1998 and 1999, so it is likely that the pool of officers remaining in 2000 and 2001 was less competitive. We are not measuring annual cohort data but how year-to-year cohort flows accumulate at a point in time. The percentages are affected by changes in both numerators and denominators that may have been influenced more by prior-year changes than current-year change. Moreover, occupational advancement rates are not independent of servicewide promotion rates. Rates of advancement for tactical operations officers may turn down, because overall rates were also lower.

Figure 3.61
Advancement of Marine Corps O-5 Officers, by Joint Experience

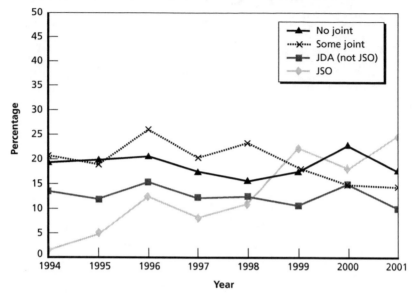

Figures 3.62 through 3.69 indicate the advancement rates for the services' tactical operations officers, by degree of joint experience. Similar to the analysis of all non–health care officers, there is not a clearly conclusive advancement pattern by joint experience.

Figure 3.62
Advancement of Army O-6 Tactical Operations Officers,
by Joint Experience

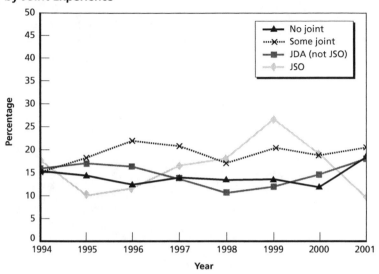

Figure 3.63
Advancement of Army O-5 Tactical Operations Officers,
by Joint Experience

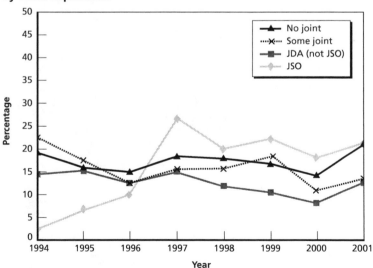

Figure 3.64
Advancement of Navy O-6 Tactical Operations Officers,
by Joint Experience

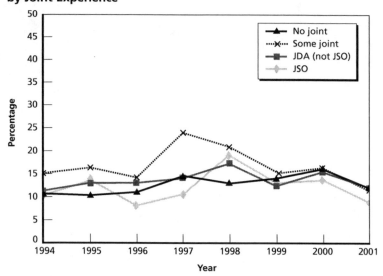

Figure 3.65
Advancement of Navy O-5 Tactical Operations Officers,
by Joint Experience

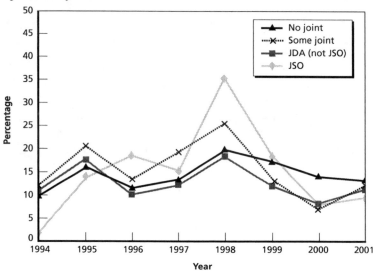

**Figure 3.66
Advancement of Air Force O-6 Tactical Operations Officers,
by Joint Experience**

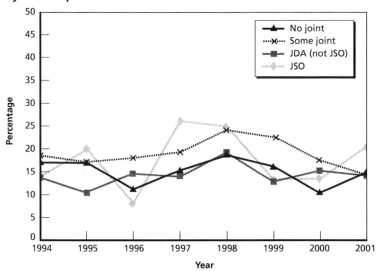

**Figure 3.67
Advancement of Air Force O-5 Tactical Operations Officers,
by Joint Experience**

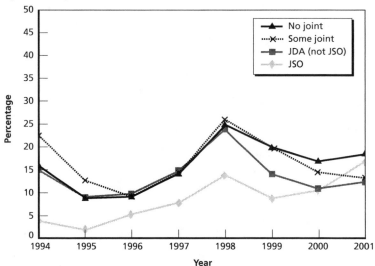

Figure 3.68
Advancement of Marine Corps O-6 Tactical Operations Officers,
by Joint Experience

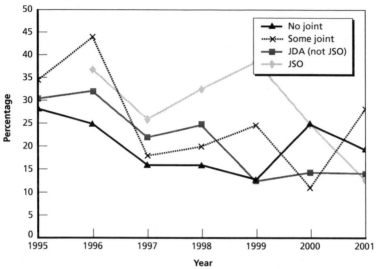

Figure 3.69
Advancement of Marine Corps O-5 Tactical Operations Officers,
by Joint Experience

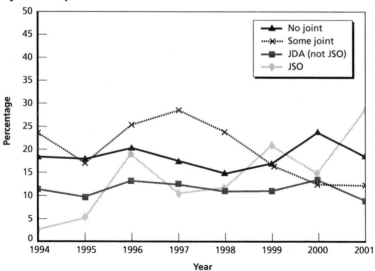

Summary

From the analysis above, we can make four general observations. First, jointness is accumulating in the officer corps, which is increasingly joint each year. This is particularly apparent at the grade of O-6 in each service. Second, the degree of jointness depends on the officers considered; certain occupations have more jointness, and certainly more need to be joint, than others. Third, jointness among the officer corps may be reaching a steady-state level. Evidence of this can be seen in this chapter's figures in which rates of positive year-over-year change are decreasing. Finally, there are some differences in management and behavior of officers with or without joint experience. In general, officers who have no joint experience advance at a lesser rate and leave at a higher rate than those who do. The differences are more apparent when considering leaving rates than when examining officer advancement, but those differences may not be significant.

Observations Based on Service Management of JDAL, Service Perspectives, and Actual Outcomes

Are the services sufficiently joint to eliminate Goldwater-Nichols? Occasionally, service personnel will assert that, based on the success of recent military missions, the resultant increased awareness of the other services, and the acknowledged integration of service missions, officers are "already joint." These "already joint" arguments are generally made by service personnel managers as a basis to diminish the future need for GNA-like constraints on officer development and promotion. Based on interviews with officer managers in all the services and with officers themselves, we conclude that although the officer corps is increasing joint, there is little supporting evidence that the services are sufficiently joint to eliminate the GNA. Nor is there, absent a strategic approach, a valid measure of how much jointness in the officer corps is sufficient. For example, while operational level missions may be increasingly joint, and officers increasingly have joint education and joint experience, service mainstream cultures do not reflect total acceptance of jointness, as described earlier. Consistent with the "check the box" mentality, we did not find a general perception of the value or importance of joint experience among personnel managers. For them, joint assignments are just some of many assignments to fill with available officers. Further, we found unanimous assent in all our interviews that any revoking of GNA-like rules would result in a return to pre-GNA practices. Additionally, the current outcomes of increasing (albeit stabilized) jointness among officers depends heavily on the assignment of the best officers to JPME II and to joint assignments, as these high-quality officers are promoted, thus permitting the officer force to accumulate, or accrete, jointness. Absent requirements for the services to share their best officers and accumulate jointness among future leaders, they would likely cease to do so. Indeed, the prediction of a likely backslide to pre-GNA practices seems to underscore the need for some rules pertaining to joint officer management, and the perception of such a likely backslide seems a very compelling counterargument to the "already joint" assertion. Moreover, it appears that increasing further the joint training and development of the officer corps requires the availability of more qualifying positions and JPME II seats, not fewer.

CHAPTER FOUR

The Joint Officer Management System:
The Need for a Strategic Approach

This chapter discusses a strategic approach in a joint context and why such an approach is needed. A strategic approach must understand the need or requirement for critical workforce characteristics as a result of missions and goals and the ability of the management system to provide officers with those characteristics. Moreover, the approach needs to demonstrate (1) a strategy or policy for aligning the availability of officers with the characteristics with the need for them or (2) a rationale for why more widespread availability of the characteristic than the immediate need for it would be desirable.

Current Goldwater-Nichols Implementation

Numerous studies of joint officer management have been done since the start of the GNA era in 1986. Many of these studies are listed in the bibliography. However, the policies chosen by the Secretary of Defense to implement the GNA have not changed considerably since the early years of the act's inception.[1]

In the Senate and House studies that preceded the act, and in the act itself, Congress, as principal, set objectives that the system of officer management put in place by the act was designed to achieve.

---

[1] See Appendix A for a discussion of changes to the 50-percent rule as the only example of significant change to GNA implementation.

These, and other objectives set by DoD, are reviewed later. The Secretary of Defense and the service secretaries were directed to implement certain provisions of law and to decide how other provisions were to be implemented. During the course of the GNA's implementation, the rules set in place by Congress or the Secretary have frequently been described as arbitrary.[2] For example, whether there should be 1,000 or 800 or any critical positions has been questioned, as has the original policy of 100 percent and 50 percent organizations.[3] "They don't trust us" seems to be DoD's perspective of Congress and the rules it imposes. From the perspective of Congress, however, such rules as it specified were needed because of incomplete information and uncertainty about how the department would actually implement the law to meet the objectives without such constraints.

Toward a Strategic Approach

If DoD could show that it is achieving the objectives set in the law, it is possible that Congress might be willing to adjust the rules dealing with critical positions, tour lengths, joint assignment and JSO requirements, and promotion comparisons. In fact, Congress has approved changes at various points in time since 1986. However, we conclude that some additional initiatives for change from DoD were unsuccessful because they lacked either context or strategy by which Congress could judge ability to meet objectives or to judge confirmation of maximal effort toward them.[4] Part of a strategic approach to

---

[2] We do not use this term pejoratively. The decisions were made through structured decisionmaking processes based on determinations about likely effects. What was arbitrary was the choice of any particular number of critical positions or percentage of jobs in an organization that provide or require joint experience.

[3] This is explained in more detail in Appendix A.

[4] For example, the Senate version of the NDAA for FY 2001 contained provisions that, among other items, would streamline designation and management of JSO and change certain promotion objectives. The House bill did not contain these provisions, and they were not adopted in conference. Also, GAO cites verbal comments from the Joint Staff that DoD

workforce planning would be to demonstrate such context and strategic intent as a basis for change or continuation of implementation practices.

Workforce Characteristics

There are three well-known requirements in law from which we can infer need for one or the other of two critical workforce characteristics: joint experience and joint education. First, the requirement for officers to have completed a JDA prior to promotion to general or flag rank sets a requirement for joint experience for most of the approximately 900 general and flag officer positions.[5] There is also a requirement to fill 800 critical positions with JSOs that sets a requirement for officers in these positions to have successfully completed JPME II and a prior JDAL assignment. Third, the requirement to fill at least half the JDAL positions with a JSO or JSO nominee sets a requirement for 50+ percent of JDAL positions to be filled with officers who have completed JPME II. (Successful completion of the assignment typically qualifies the officer as a JSO, thus meeting the requirement in law.) On the other side, there are constraints either in law or in DoD policy that affect the availability of officers with the joint characteristics. For example, qualifying joint experience can only be obtained in billets external to the military service. These billets for obtaining qualification are further limited in that they must be in grade O-4 and above and only some of the billets in defense agencies can provide the qualification.[6]

Need for and Availability of Characteristics

The following two notional diagrams portray the contrast for need and availability between the current system and a system premised on

---

"views provisions in the act as impediments that must be removed before it can develop an effective strategic plan." Conversely, GAO states that DoD "will not be able to demonstrate that changes to the law are needed unless it first develops a strategic plan" (GAO, 2002, pp. 31–32).

[5] Some officers, such as doctors, are exempt from this requirement.

[6] Appendix A explains the qualification details.

a strategic approach. In the current system, as shown in Figure 4.1, those billets that have a prerequisite need for joint experience are largely a subset of those that provide joint experience. (Diagrams that portray the need for JPME II would be similar.) Much of the emphasis of Goldwater-Nichols and the DoD implementation has been on identifying positions that provide a valid joint experience given that officers serve in them for a minimum amount of time. All such positions (identified by the large circle) must be external to or outside the military service. The key component of a strategic approach, the need for officers with such experience, is shown by the two small circles. Joint experience is needed in 800 critical billets that are all outside the service and in most of the 900 general and flag officer billets, many of which are internal to the service.

As shown in Figure 4.2, a strategic approach involves a deliberate determination of which jobs, inside or outside the service, need joint experience or provide it. Specifically, given the current strategic intent of DoD with respect to jointness ("push it to its lowest appropriate level"), the need for joint experience should be measurable in a much larger number of billets, particularly in billets internal to the service. Moreover, valid joint experience might now be provided by

Figure 4.1
In the Current System, the Need for Joint Experience Is a Subset of Availability

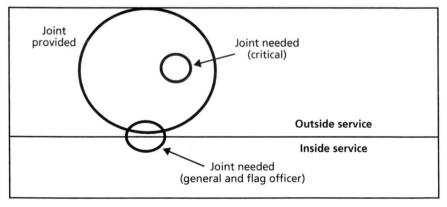

Figure 4.2
In the Strategic Approach, the Need for and Availability of Joint Experience Is Determined

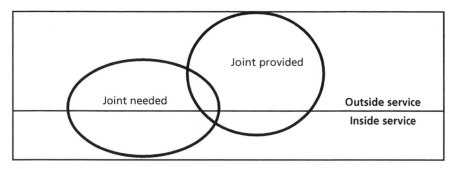

service in billets internal to the service, particularly those associated with joint task forces, with service component commands and with joint planning and operations. The key components of a strategic approach can be discerned from Figure 4.2: Which jobs require or provide joint experience? How many of each exist (what is the size of the two circles)? What is needed to align the two circles?

Strategic Approach as Basis for Change

Alternatives for change suggested over the years by DoD tend to deny that a need for the characteristic of joint experience or education exists or tend to increase the availability of the characteristic. For example, eliminating the concept of critical positions or reducing the number of them reduces the need for joint experience and education. Allowing in-service billets to be qualifying for providing the characteristic of joint experience or reducing tour length to achieve qualification increases the availability of officers with the joint experience characteristic. Such alternatives for change appear to lack context or strategic intent because they do not seem to match availability of joint experience or education to the need for them; appear to reduce the value of "jointness" (little need and excess availability) at a time when most observers are asserting an increased value for it; and appear to fly in the face of congressional intent and DoD rhetoric on jointness, especially looking toward the future.

Why a Strategic Approach?[7]

Apart from how to take a strategic approach, why should an organization take a strategic approach? The basic reasons are that (1) human resource management in an organization has multiple influences that are often in conflict and that (2) primary actors within an organization with different values and attitudes have leeway to make human resource choices. A strategic approach determines which influences are more important and limits choices of the actors to those most conducive to organizational performance.

Influences

Three organizational influences affect human resource management. The first is the organization's administrative heritage, which includes structures, methods, and competencies of the past. For the military, it is the lingering legacies of the Department of the Navy and the Department of War. The military services have long memories with respect to how human resource management has been done. The services know how to develop, educate, and train, assign, and promote officers. They have been doing it for two centuries and have been operating the basic design of their systems since at least the end of World War II with the passage of the Officer Personnel Act of 1947. The Grade Limitation Act of 1955 and the Defense Officer Personnel Management Act of 1980 extended these basic designs; the GNA disrupts them by mandating that different career paths and developmental practices are needed.

---

[7] This section is based on J. Paauwe and P. Boselie, *Challenging (Strategic) Human Resource Management Theory: Integration of Resource-Based Approaches and New Institutionalism*, Rotterdam, The Netherlands: Erasmus Institute of Management, ERS-2002-40-ORG, April 2002, and authors cited therein.

The second influence is culture and law. This influence imposes currently prevailing values and norms such as fairness, equity, merit, and equality of opportunity to establish relationships with both internal and external stakeholders. Officership is the issue, and it has been service based. The military operates within a closed system and a fundamentally different legal paradigm than that of the private sector, with different notions of fairness, equity, and merit. Age and other discriminations are practiced in one in ways that are illegal in the other. The Fair Labor Standards Act and the Employee Retirement and Income Security Act are among the nation's laws that do not apply to the military, which has its own legal and social practices as a result of law, executive order, or policy. The GNA disrupts this in that it suggests that some officers are more valued than others in a way that the current system has not accepted.

The third influence is the mission and technology orientation by which national security is produced and delivered. The issues here are efficiency, effectiveness, flexibility, quality, and innovativeness. GNA disrupted this in significant ways by bringing combatant commanders and their needs to the forefront. The services and DoD have adjusted to and have even embraced this aspect of the GNA on the operational side but not the management side.

Constraining Action While Providing Flexibility

There are multiple decisionmakers in the human resource system. Among them are Congress, agencies in the executive branch, the Secretary of Defense and his staff, the CJCS, the military departments and the chiefs of service (and their respective staffs), organizations that use officers, and officers themselves. Constraints are needed on the choices that these decisionmakers could make within an unconstrained system; however, the system should not be overly constrained such that it becomes inflexible. Decisions to be made are about "fit." The human resource system must have strategic fit in that it supports the strategy of the military and of its organizations. It must be based in the need to be successful in prosecuting military operations and delivering national security. The human resource system must also have organizational fit in that it must work in conjunction with other organizational and administrative systems such as the deployment and readiness systems, which are themselves changing. The human resource system must have environmental fit. The strategies used must be in consonance with the practices and norms of the larger external environment and with the needs of prospective and serving officers. And last, the system must have internal fit. Human resource practices must be coherent and consistent bundles of policies and practices.

Shaping Human Resource Strategies Toward Organizational Performance

Military operations and organizational fitness for them requires shaping human resource strategies to generate outcomes that contribute to performance of the organization. In essence, the missions and goals are emphasized, while other influences such as administrative heritage and cultural norms are recognized. By suggesting a new competitive strategy for the military that leads to changes in organizational and administrative systems and cultures, the GNA has imposed different constraints on the multiple decisionmakers in the system that require adjustments. A strategic approach to human resource management is important to achieving organizational goals and missions.

Summary

The need for officers with joint experience and education connects directly to DoD goals and missions through organizations that prosecute those goals and missions. Determining the need for officers with these qualifications enables DoD to answer such questions as to how many JSOs are needed and how many JPME II school seats are needed.[8] Moreover, if the need for joint experience and education in the officer corps cannot be met with existing management practices,

---

[8] GAO has asserted that these are key questions to be answered for the active component in determining a strategic approach to the development of officers in joint matters.

the gap can serve as a basis for changing management and controls by either DoD policy or changes to Title 10. Conversely, however, if there is little identified need for officers with joint experience and education, then the means for accomplishing expressed DoD missions and goals through joint operations, and even the goals themselves, could be in question.

CHAPTER FIVE

How to Implement a Strategic Approach for Joint Officer Management

A strategic approach determines the need for critical workforce characteristic(s) given missions, goals, and desired organizational outcomes; assesses availability of the characteristic(s) now and in the future; and suggests changes to human resource management for personnel with the characteristic(s) to minimize gaps.

Our approach benefits from and reflects GAO's guidance regarding strategic approaches as well as prior RAND research employing that method.[1] Several RAND studies have taken a strategic approach. Thie and Brown's 1994 report, *Future Career Management Systems for U.S. Military Officers*, responded to congressional direction for a study of officer management determined requirements for officers (by service, grade, and occupation) as a function of national military strategy, organizational design and structure, force size and active-reserve component force mix, doctrine and operational concepts, and technology. The study then applied career management principles to derive alternative systems for developing and managing officers to meet these needs.

A later report (*Interagency and International Assignments and Officer Career Management*, Thie, Harrell, and Emmerichs, 1999), which was also in response to a congressional requirement, identified interagency and international positions and designed career models

---

[1] General Accounting Office, *Human Capital: Key Principles for Effective Strategic Workforce Planning*, GAO-04-39, 2004.

and judged the feasibility and desirability of using particular models to develop and manage officers to fill these positions. These career models are more fully discussed later in the report.

Most recently, RAND published two related studies by Emmerichs, Marcum, and Robbert (*An Operational Process for Workforce Planning*, 2004a; *An Executive Perspective on Workforce Planning*, 2004b) that demonstrate how to apply a strategic approach for the acquisition workforce and demonstrated it in certain organizations. These studies identified the essence of the approach in four steps:

1. Identify critical characteristics an organization needs to carry out its intent.
2. Assess current workforce availability of critical characteristics.
3. Estimate availability of critical characteristics in the future.
4. Develop policies to eliminate gaps.

Moreover, applying these steps in organizations led the authors to conclude that there are three key factors to success in implementing a strategic approach:

- enthusiastic executive and line participation
- accurate and relevant data
- sophisticated and comprehensive workforce projection models.

We briefly discuss GAO's steps to a strategic approach here and more fully describe its component implementation parts of data collection, data analysis, and decisionmaking in the next section. Our recommended approach has five major steps:

1. Determine workforce characteristics that will be needed in the future to meet strategic intent. We believe that these characteristics can be aggregated into proxy variables for competencies based on experiences such as joint multiservice, joint interagency, and

joint multinational[2] and on joint education and/or joint training as discussed later in this section. Analysis of survey responses could eliminate one or more of these or lead to determination of need for other characteristics. The accuracy of billet needs with respect to characteristics such as grade (experience), occupation, and other characteristics will need to be assumed.

2. Determine needs for these characteristics of joint experience, education, and training. Where (in what positions) are officers with joint experience, education, and training needed? How many of these positions are there? Does this differ across services, for different occupations, or at different levels of seniority? Does the need for such officers extend to in-service billets? For this step, we recommend collecting data as to which external and in-service billets need an officer with prior joint experience or education.

3. Identify officers with these characteristics who are currently available. We recommend using existing personnel databases to assess the current numbers of officers with the experience and education characteristic of interest. We recommend surveying all external billets (and selected in-service billets) to determine those billets that provide joint experience to officers as a basis for projecting future availability. Current numbers and timing of JPME II seats[3] will likely need to be used as the start point for projecting educational qualifications.

4. Use models to

a. project availability of officers with these characteristics in the future, given certain career management practices

---

[2] We have begun to see use of the acronym "JIM" to reflect joint (multiservice), interagency, and multinational as separate components of the larger concept of jointness. This will be discussed further in this chapter's section on data collection.

[3] As of this writing, the House Armed Services Committee has a provision in its version of the 2006 NDAA that would increase the number of institutions that could provide JPME II and thus increase the number of JPME II graduate rates. Depending on the final outcome of this provision, we will incorporate any changes in the modeling approach we take in the next phase of our research.

  b. calculate future gaps between the need for officers and the availability of them[4]

  c. refine and evaluate near-term policy alternatives to reduce gaps within the strategic context

  d. develop strategies that address long-term issues for reducing the gaps.

5. Identify other implications of the strategic approach such as effects on objectives and desired metrics for evaluation.

    In summary, the strategic approach needs to confirm the characteristics of interest. We assert they are multiservice, multinational, and interagency experience as well as joint training and joint education. Data need to be gathered regarding the need for jointness, the current stock of jointness, and the possible future provision of jointness. Modeling should confirm the degree to which jointness can be accreted and the extent to which the future stock of jointness will satisfy the future identified demand for jointness.

    The following sections provide detail on workforce characteristics, data collection, and analysis to implement the steps in this strategic approach.

Workforce Characteristics

The officer workforce is traditionally specified in terms of grade and length of service (experience) and occupation (knowledge areas and skills). A further research effort would need to assume that positions are correctly specified and that officers are correctly designated in these terms. Four additional characteristics are of interest in our

---

[4] We are using the logic that underlies strategic human capital management of matching availability of workforce characteristics to the demand for them. This assumes that there is a cost for developing people with these characteristics so that both an over- and undersupply of the characteristic are not desirable. However, other assumptions could be made that change the nature of the assessments we are making. For example, the availability of officers with joint experience and/or education could lead to increasing demand for them in many military positions. The availability of such officers could by itself create a need for them.

study: joint experience, joint education, joint training, and joint acculturation. One can explicitly measure the first three. Joint acculturation—values, attitudes, and beliefs about jointness—is obtained in many ways, such as through attendance at schools with officers of other services, reading and self-learning, planning and operational experiences, and exposure to officers of other services, agencies, and nations. One can treat joint acculturation as an outcome of the other three characteristics, although we recognize it could be obtained separately from them.

Each of these characteristics exists in at least three "flavors": multiservice, multinational, and interagency,[5] and one can measure the characteristics on that basis. Moreover, these characteristics exist on a low-to-high continuum governed by such factors as organizational level (e.g., national vs. battalion/squadron, strategic vs. tactical), importance of function performed (e.g., planning vs. protocol), frequency (how often performed), duration (how long performed), and intensity (conditions—e.g., wartime 12-hour shift work in theater vs. peacetime routine staff work in the United States). For some positions, these factors are relatively permanent and inherent in the billet. For other billets, the factors, particularly intensity, might be situational. We recommend a further effort to collect data about these and other factors for each characteristic in each of its flavors.

Need for Workforce Characteristics: Data Collection

To implement a strategic approach, data are required about which billets (those in service organizations as well as those external to the service) require the officers filling those positions to have some prerequisite degree of jointness. This can be expressed in discrete catego-

---

[5] We will measure the characteristics in these terms. One alternatives would be to measure in the aggregate, that is, to subsume these three in the term "joint" as is currently done. Another alternative would be to further disaggregate these elements into knowledge areas and skills appropriate to each. Such disaggregation is not needed for our study, and aggregating into joint eliminates important distinctions that might be analytically and policymaking useful.

ries such as "required," "desired," or "not needed" for several different elements of jointness, including multiservice (MS) experience, interagency (IA) experience, multinational (MN) experience, job-specific joint training (such as that concerning specific systems or processes used), and more general joint education. These different elements of jointness are important for addressing proposals either to limit the definition of what constitutes jointness or to readdress the education and training provided for joint officers.

Collecting data about billets that have a need for jointness can be managed through existing manpower or personnel offices in organizations external to the services and in service organizations. The decisions about billets could be made at the directorate level within organizations external to the services for billets that they control.[6] Regarding billets within the services, such determinations might involve a central manpower capability, such as the G-1/3 of the Army for certain billets (e.g., O-6/O-5 commands, service headquarters planning staffs) and major commands for other billets (service component command staffs[7] or other positions that require or would benefit from prior jointness). The determination should be made at the appropriate level to ensure ownership of the results.

We describe here a one-time data collection effort for use in this research.[8]

---

[6] In our interviews with unified command staff, this was routinely viewed as the best place to make such a determination. The manpower office (J1) would manage the process; the various directors would make the determination; and the results would be approved by the Chief of Staff or Deputy Commander.

[7] As above, our interviews suggested a manpower/personnel-managed process, determination by managers, and approval by a chief of staff or deputy commander would be the best process.

[8] As per guidance received from the research sponsor, this report includes instructions for a single collection of data, which was judged to be more expedient and thus more supportive of a conversion to a strategic approach. Longer-term maintenance of a strategic system will require the services to collect and maintain joint-specific information on their officers' qualifications, and the services and the joint staff to collect and maintain joint-specific information regarding their organizations' billets. Appendix B provides additional discussion on such long-term data issues.

Each billet in the target group (O-3 through O-10) needs to be assessed for two reasons. First, in advance, no determination can be made that all billets in an organization require joint education or experience. Second, other information about the billet, such as grade and occupation, would need to be extracted. However, during the determination, a service might use general rules (e.g., all O-6 commands require prior joint experience) in which case the subsequent analysis can then take the grade and occupational composition from the rule.

Data should be collected on all billets currently on the JDAL, on all non-JDAL billets in organizations external to the service, and on selected billets in the military services. The research sponsor should assume the role of informing the services and external organizations about data-collection procedures and ensuring timely participation. Appendix C contains draft memos that could be used to explain the data-collection procedures to involved organizations, while Appendix D includes a draft memo of instruction for distribution to individual points of contact at each of the organizations involved in the survey.

The Office of the Secretary of Defense and the Joint Staff have a role encouraging the services and other organizations to evaluate each billet with a fresh eye toward the strategic context of missions and goals and consider whether, for example, O-6 commanders require prior joint experience or whether they would benefit from it or perform better having had it. Such determination should flow from the operational consideration of jointness that has been embraced by the national security establishment. This is a significantly different consideration from calculating after the fact whether O-6 commanders have had previous joint experience.[9]

---

[9] For example, 39 percent of Army FY 2001 combat arms tactical brigade (O-6) commanders had a prior joint assignment, and 42 percent of them had completed JPME II (David E. Johnson, "Preparing Potential Senior Army Leaders for the Future: An Assessment of Leader Development Efforts in the Post–Cold War Era," Santa Monica, Calif.: RAND Corporation, IP-224-A, 2002). The question here is whether such experience is required or desirable prior to serving in such positions. In the views of senior officers we interviewed in combatant commands, the operational environment of blending multiservice, interagency, and multinational capabilities at the O-6 command level is too critical to allow it to be learned ad hoc and "on the fly."

The process of collecting data to determine billets that require or would benefit from prior jointness will require each service and external organization to indicate for each O-4 through O-6 billet whether the following kinds of experiences were required, desired, or neither required nor desired: MS, MN, IA, job-specific joint training, or more general joint education. Thus, at the conclusion of this exercise, one would understand for all existing O-3 through O-6 billets, regardless of where they are, the need for prior jointness. Further, the analysis could tie that data input to billet descriptors to characterize the types of billets that have such needs and connect this need to the alternative side of our analysis: the provision of officers with joint experience.

Different Aspects of Jointness:
Education, Training, Experience, Acculturation

Goldwater-Nichols refers only to education and experience as methods of developing joint officers, but we suggest that educating, experience, training, and acculturation are separate components that should be considered as such, and have included them in workforce characteristics for this strategic approach. Much of the confusion about the purpose and value of joint officer education, as well as the debate about where jointness can be acquired,[10] is more understandable when these four elements are considered separately. For example, it is possible that the GNA provision that joint education must be provided at a single, mixed-service location was in some part due to the concern that students of different services would be intermixed and thus that students would receive some acculturation. The requirement that only billets outside the services can be included on the JDAL and provide joint credit is likely some combination regarding concern over which billets provide a valid joint experience and the expectation that officers should be surrounded, and thus acculturated, to other services and experiences.

---

[10] Currently, officers can only receive joint credit while serving in a JDA, which by law must be in a billet outside the services.

Similarly, there is some confusion and lack of agreement about the value of JPME II compared with the investment required for officers to attend JPME II. This is likely a reflection of the confusion between education and training. In other words, some officers, such as those who will use the Joint Operation Planning and Execution System (JOPES) in the course of their work, are receiving job-specific training when they attend JPME II and are thus more positive about the value of JPME II to their work. Other officers are receiving general education, not job-specific training, since the content of JPME II reflects their work responsibilities only indirectly. These officers are likely to be less positive about the job-specific impact, even if they valued the education and acculturation. Furthermore, to the extent that officers are receiving JPME II after they have already begun their joint assignment and thus have already oriented and self-trained themselves, there is limited return to the time invested in JPME II, regardless of whether their curriculum provides them with job-related training.

This research did not evaluate the curriculum of JPME II. However, we urge future curriculum reviews to consider the difference between education and training and to determine the purpose of JPME II. To the extent that the JPME II curriculum is intended to provide education, such education improves the quality of the officer force overall and is a valid investment of the services. To the extent that the intent of the JPME II curriculum is to provide job-specific training, such training affects the performance of officers at joint organizations and is a valid investment of those organizations. To the extent that the intent of JPME II curriculum is to provide acculturation, then the mixed faculty and mixed student body are extremely important. However, acculturation can also be provided through a joint duty assignment.

Availability of Workforce Characteristics: Data Collection

Determining Individual Billets That Provide Valid Joint Experience

Officers currently receive joint credit for serving in positions on the JDAL. One frequent criticism of the current system is that officers are serving in other assignments that provide a rich joint experience but do not grant the officer joint credit. Likewise, there are officers serving in assignments on the JDAL that may not provide what some would consider a joint experience, either because of the content of their work or because of limited interaction with other services, nations, or agencies. Without yet establishing the criteria by which an assignment can be considered to grant a "valid" joint experience,[11] the issue at hand is whether it is important to only grant credit to those who have been judged, with forthcoming criteria, to have gained a valid joint experience. This recognizes that there are differing degrees of jointness; billets can be ranked from high to low validity for the joint experience they provide. A cutoff score could be determined at the point in a ranking of all billets at which the need for the characteristics of joint experience and education was met by managing officers through sufficient positions to achieve it. Alternatively, a fixed cutoff point based on validity of joint experience could be set that might lead to the need being met or not met.

These new strategies would constitute a considerable change to the current system, evidenced by the number of current policies that would be forced to change. For example, there would likely be no "100 percent organizations" in which all serving officers in grades O-4 through O-6 receive joint credit. In our 1996 study of these issues, we specified an algorithm that defined a valid joint experience and applied it billet by billet. For a JDAL of about the same size as today's, we found that the percentage of billets that provide jointness were as follows: Joint Staff (88 percent), combatant commands (89 percent), other unified commands (76 percent), Office of the Secretary of Defense (81 percent), defense agencies (71 percent), and all

---

[11] Criteria would include some combination of variables such as frequency of multiservice, interagency, or multinational interactions and nature of duties or functions performed.

others (83 percent). We believe that any billet-by-billet analysis would lead to similar conclusions: The current 100 percent organizations are in reality less than that. Moreover, all organizations appear to be closer to 100 percent than to 0 percent. While there would be split organizations, with some officers receiving credit and others not, officers performing similar tasks and interactions would receive joint duty credit if applicable; there would not be "haves" and "have nots" with similar duties. Our study also recognized that almost all external billets had some joint content and that expanding credit to all external positions has advantages and disadvantages but greatly simplifies implementation.

We describe a process under the assumption that it is important to grant credit only for a "valid" joint experience after a billet-by-billet analysis.[12] We recommend the use of these billet-level data to assess the degree of jointness the billet provides to officers (most likely a continuous variable derived from multiple criteria such as amount or level of interaction and duties). This measure is inherently more complex than the discrete measure used to assess whether prior jointness is beneficial because, to the extent that criteria are used to judge the validity of joint experience, disaggregated data about those criteria must be collected. The sole method that would confirm that only officers with a valid joint experience (absent a very broad definition of "valid joint experience") receive joint credit is a billet-based inclusion process. The billet-based analysis is necessary to identify those billets providing a joint experience, whether all such jobs could provide joint credit, or whether they could be prioritized for a list capped in size.

Types of Data Needed

Specific data needs contributed to the construction of a recommended data collection instrument (Appendix E). We reviewed recent studies that collected data regarding joint billets to use in

---

[12] If, however, it was known in advance that certain characteristics of the current implementation would be kept, then the size of the data collection effort could be reduced, as fewer billets would need to be assessed on a billet-by-billet basis. We discuss these ramifications later.

workforce planning studies. We feel that, although each survey asked relevant questions, no one particular survey captured the entire picture necessary for a thorough analysis. We accomplished this goal by continually referring to these other surveys as we built our own, which ensured we had not omitted important questions or answer choices.

We can categorize needed data as either objective or subjective and closed-ended or open-ended. The first type of required data falls into the closed-ended, objective category, which is information such as organization, billet number, grade, occupation, branch of service, and similar information. Most of these data already exist in a billet file, but a survey may ask for these data as a means to verify the billet data and to gather the data for positions not included in the JDAL. We researched several prior surveys and found that there are numerous ways to represent these types of data. For instance, some surveys asked for different variations of numerical job designation (e.g., air force specialty code [AFSC] or military occupational specialty [MOS]), while others asked only for the individual's title—of which there were many alternative ways of expressing.[13] With regard to education, most simply asked whether or not the respondent had completed JPME, some included questions about other types of military or civilian education, and a few asked more in-depth questions about the completion of JPME. We also included some closed-ended question dealing with characteristics of the person holding the job, such as educational background.

The next category of questions is more subjective, but it examines in more depth, with closed-ended questions, the day-to-day responsibilities and duties associated with a billet. These questions address duties; time spent on joint matters; whether such time was MS, IA, or MN; and interaction with other organizations. Through-

---

[13] The advantage of asking for a respondent's numerical designation is that it eliminates any ambiguity regarding the individual's actual job, and it is less complex to process. However, when using just the numerical designation, errors in both entry and transcription must be anticipated. Including the title with the AFSC, MOS, branch code, or Navy designator can serve as a means of verifying accuracy.

out the surveys we researched, we found that the boundaries of duties and subject areas are not always clear. For instance, across the surveys the "functional" area encompassed duties related to finance, planning and logistics, science and engineering, human resources, and even command in different ways. In particular, for this study it is vital to know the context in which duties are performed (own service, other service, other nation, other agency) and the amount of time in that context. We also recommend understanding the intensity and the deployment status of the job as part of the context in which duties are performed.

We also included some closed-ended, subjective questions about various aspects of jointness or career development. Respondents would likely not be required to answer these types of questions. The recommended survey instrument includes only one open-ended question at the conclusion, permitting participants to provide comments or additional thoughts relevant to the issue.

Use of Criteria to Determine a Valid Joint Experience

There are two aspects to determining the criteria by which to evaluate individual billets.

First, one must establish which criteria to consider, such as job content or responsibility and job context or number of services, agencies, nations involved on a daily basis. Prior RAND research developed an algorithm that consisted of two criteria. The first was time spent in joint (the proportion of a billet's time spent on matters involving other services, agencies, and nations). The second was a weighted assessment of job function (the "job" of the billet, including areas of work and duties to be performed). We eliminated other possible criteria because they led to an over-specified (too complex) answer that was not significantly different from the simpler algorithm. We combined the criteria into one score for each billet. Since that study, other criteria have emerged as potentially useful. For example, officers currently receive joint credit if they serve in identified JDAs for the minimum assignment length (tenure in job). Should the tenure required for credit vary depending on the intensity of the position or the degree of jointness of the position? In other words, if someone is deployed and serving in a joint billet during a time of conflict so that their experience is intensely joint, should they be considered experienced to a requisite level to receive full credit for half as long an assignment? Said another way, should they receive twice as much credit for their experience, or should they just receive "partial" credit representing the duration of the time they served in a joint position? Alternatively, given that some jobs provide less joint-ness because they are at the lower end of a joint continuum or are only "associated with joint matters," should such jobs require longer tenure before providing joint credit or acquire partial credit at a slower rate? We recommend collecting data on multiple criteria that could be used to determine jointness. Moreover, it is entirely possible that multiple algorithms might be created. One algorithm could determine valid joint experience for external positions, and a different algorithm might be used for in-service billets.

Second, one must determine how joint a job must be for inclusion. In other words, if jobs are evaluated by a function of prioritized criteria, then one could develop a continuum of jobs in which those at one end are extremely joint and those at the other end are mini-mally joint. The question then becomes: To what degree of jointness do jobs need to satisfy to be included on the JDAL? Other research[14] developed three categories of jointness: positions critically related to joint matters, positions directly related to joint matters, and positions associated with joint matters. This schema would suggest that either all positions are included, that is, are joint enough if they have some joint content, or that positions only associated with joint matters are deleted. A continuum of jointness would permit a decision to cut inclusion at any point but would still require a policy decision of how joint is sufficiently joint. Such decisions in the past have been pri-marily subjective. Going forward, these decisions can also be informed by the need for jointness determined in this approach.

---

[14] Booz Allen Hamilton (2003).

Managing Data Collection

Data could be collected through a Web-based census of all positions currently on the JDAL, of non-JDAL positions in external organizations, and of in-service positions nominated by the services. The census instrument would be directed at the incumbent of the position, with the supervisor or another knowledgeable individual as the backup. The sponsor of the research would be considerably involved in soliciting nominee billets and ensuring near 100 percent completion of the instrument. We include sample census protocols in Appendix E.

Projecting Availability and Calculating Gaps: Data Analysis and Modeling

The best data analysis for this effort would consist of two different and complementary kinds of analysis. First, one should conduct a systemic analysis to determine the feasibility of matching the availability of jointness to the need for it, given an established need for joint experience, a rate at which officers with joint experience are developed, constraints such as limited educational opportunities, policy and practices, and likely behavior of officers in the system. This analysis should be based on three different kinds of inputs: (1) the data regarding the need for officers with joint experience, (2) the number and types of billets that provide officers with joint experience, and (3) the management frameworks that capture the policies of managing such officers and predict the individual behaviors of officers with joint experience. These sets of inputs are described in more detail below. The systemic analysis could provide a general feasibility assessment of each alternative (scenario and framework combination) for each service overall, for the line communities of each service, and for selected occupations (e.g., Army infantry, special operation forces [SOF], and intelligence). This feasibility assessment could highlight whether the percentage of officers with joint experience within a given service, community, and pay grade are sufficient to fill the billets that demand such officers. One could also ascertain whether the number of billets that provide jointness, the management method, or the number and timing of school seats are the constraints to feasibility. If, for example, all Army SOF O-6s have joint experience, then the percentage of O-6 SOF billets that require such experience is irrelevant; the system is clearly feasible. If only 30 percent of such officers have joint experience and 40 percent of the billets require such experience, then the system is clearly infeasible. If, however, 60 percent of officers have joint experience and 50 percent of the billets require it, then that is more difficult to assess. Do the right officers have joint experience? Is there sufficient assignment flexibility? In this case, while one could explore policies that increase the amount of jointness in a community, it may be difficult to assess feasibility conclusively. Thus is the need for the second analytical process.

The second kind of analysis would also conduct career path analysis for a limited number of occupations and would focus the analysis on selected line communities, such as surface warfare officers and infantry officers. This analysis would allow one to understand the implications, within the expected career path, of inserting joint assignments. The analysis should assess how the timing of joint assignments interfaces with other key assignments for their career progressions, such as command opportunities. The analysis should also determine how changing assignment length, promotion timing, or other policies would affect the ease with which such assignments fit into a career path.

Management Frameworks

The policy analysis should interface with modeling and suggest policy implications based on analytical results. At the heart of the policy analysis are management frameworks that can be used to link management practices more closely to objectives for joint officer management and thus understand how different practices might apply to different groups of officers.

We describe below four different management frameworks, developed in prior RAND research.[15] These frameworks provide input to the analysis in the form of policy decisions, management choices, and anticipated individual behavior.

Managing Leader Succession is a career management system that, in this context, would emphasize providing future leaders with joint experience. This system would feature relatively shorter joint assignments, consistent with a pattern of developing officers. It would also suggest relatively higher promotion rates of officers who served in joint assignments, higher retention rates of those officers, and a greater likelihood that general and flag officers had gained joint experience. This system is likely most appropriate to the service line communities, from which future leaders are being developed.

A **Managing Competencies** system would place emphasis on developing intensely experienced officers in joint matters and would result in something that might be considered a joint cadre. Essentially, officers who served in a joint assignment would be highly likely to serve repeatedly in joint assignments, and they could serve in longer, more stable joint assignments. While fewer officers overall would have exposure to jointness, those who did would have very deep joint experience. This joint experience, however, could come at the cost of maintaining a service expertise, depending on the nature of the officers' occupational specialty. These officers would likely retain at a rate roughly equal to their field-grade peers, but would not have the same opportunity to be promoted to general and flag officer ranks. This system is likely most appropriate to occupations that are already highly joint, such as special operations, in which officers complete repeated joint tours but maintain their occupational expertise because their occupation is inherently joint.

A **Managing Skills** system would be designed to distribute joint experience throughout the officer corps. In this framework, there would be less emphasis on exposing the highest quality officers to

---

[15] Harry J. Thie, Margaret C. Harrell, and Robert M. Emmerichs, *Interagency and International Assignments and Officer Career Management*, Santa Monica, Calif.: RAND Corporation, MR-1116-OSD, 1999.

joint experiences, and more emphasis on maximizing the number of officers who have joint experience. Given that premise, there would be more attrition of officers with joint experience and, depending on the number of assignments that could provide a joint experience, potentially less likelihood that senior officers would have received such experience. While a managing leader succession model attempts to send high-quality officers to joint assignments and retain and promote them at a high rate, this model sends more average officers to joint assignments and then retains and promotes them at the average rate, so that the senior leadership may eventually not be as highly representative of officers with joint experience. This system could be appropriate to any service that is reluctant to identify likely future leaders or, in the case of the Marine Corps, places emphasis on the equity of the system. The potential downfall of this system, if implemented for line officers, is that it would make it difficult for the services to satisfy systematically the legal requirements for joint experience and education by promotion to general or flag officer rank.

Managing the Exception systems focus on the positions and manage assignments rather than officers. The focus of this system is "Who is available when we need someone?" This has been the historic method of managing joint assignments and does not systematically develop joint experience, provide the appropriate people to joint organizations, or invest in the future leadership. Indeed, it is difficult to identify a service or occupation that would be best served by a management system that did not focus on developing officers.

We note that a single management framework may not apply to all communities in all services. Instead, there may be a mixture of frameworks, such that different communities of officers are managed differently. Because the frameworks also reflect the objectives of the system, it is also possible that different objectives are achievable by different communities of officers. For example, the manner in which line officers are managed will impact the likelihood of accomplishing the objective to provide military leaders with joint experience. However, those may not be the same officers that serve maximum-length joint assignments and provide stability in joint organizations. Our intent is to determine the best way to manage officers to achieve a feasible strategic approach to joint officer management and to then determine which objectives are achievable with which groups of officers, which will serve as input to eventual policy decisions.

Identifying Policy Implications

Using the data about need for and availability of the workforce characteristics and the management frameworks, the analysis should inform policymakers and provide input to such policy issues as the following:

- Which variables should be included in a definition of qualifying joint experience?
- How "joint" should a billet be to be considered validly joint?
- Should MS, MN, and IA be managed or tracked separately? Are the needs and sources of each distinctly separable?
- Should minimizing oversight and repetitive measurement be a consideration? For example, if an organization is 95 percent joint, should all billets at that organization receive joint credit? If all billets are somewhat joint, should all billets provide joint credit?
- What management frameworks are suggested for different services and occupations? How different is that from the status quo?
- Are there other occupational considerations? For example, will some career fields have more difficulty gaining jointness?
- What is the relationship between necessary education and training and existing resources? Are more resources (seats) needed?
- What should be the objectives and metrics of a system to develop officers in joint matters? This is an important question to answer.

Additionally, the appropriate application of objectives will emerge with the management framework selected for different populations of officers. For example, should the services choose to manage their line officers, or future leaders, with a managing leader succession framework, then future leaders would gain joint experiences, and rich service tactical and operational expertise would be available to joint organizations; however, that population of officers is less likely to have considerable prior joint experience and will not provide the stability of long tenure to joint organizations. On the other hand, if selected occupations such as special operations, intelligence, or communications are managed within a managing competencies framework, then they will provide greater stability and prior joint expertise to joint organizations, although assignments for these kinds of officer are less likely to promote joint exposure for the future military leaders.

CHAPTER SIX

Conclusions and Recommendations

Goldwater-Nichols deserves some reconsideration, given the increasing number of joint operations, the recognition of the value of jointness among officers, and the changing management practices for officers with joint experience. All the original objectives of the GNA may not still be appropriate, and considerable conflict exists within the GNA objectives as well as between the GNA objectives and the stated goals of the services, the joint organizations, and individual officers.

However, it is not clear that the types of constraints and requirements stated in the GNA should be eliminated. Military missions are increasingly integrated, and military officers are increasingly joint. However, some cultural resistance to officers' jointness still remains. In other words, the officer management systems in most of the services are still generally resistant to developing joint officers and would likely revert to management processes that did not support jointness in the absence of GNA-type requirements, constraints, and reporting mechanisms.

A strategic approach to joint officer management, as outlined here, aligns human capital with the organization's mission, rather than empowering other influences, such as organizational, administrative, and cultural heritage or the current social, cultural, and legal practices and beliefs. The strategic approach described herein for joint officer management considers and balances the assignments that require joint experience, education, training, or acculturation with the ways officers receive joint experience, education, training, or acculturation.

The next research step is to operationalize or implement the strategic plan for joint officer management. This implementation will require extensive data gathering and complex modeling and data analysis to formulate appropriate policy alternatives.

The data required will include (1) data pertaining to positions throughout the services as well as in joint organizations that require prerequisite joint experience, training, or education and (2) data pertaining to positions that provide valid joint experience. The positions that provide a valid joint experience could include positions of any pay grade within joint organizations or in service organizations. The subsequent analysis would determine the extent to which the need for officers with joint experience can be satisfied by the number of billets that provide joint experience combined with the assignment, promotion, and management practices for officers of different communities. A strategic plan should determine the policies and practices to align the needs for jointness with the amount of jointness available among officers, given different management practices. This report provides the implementable means to do this.

A Primer to the Goldwater-Nichols Act

How Goldwater-Nichols Came About

Congress recognized the need to enact broad defense reorganization and reforms in the early 1980s. The need for Congress to act was spurned by a series of operational challenges and failures that could be attributed to the inability of the services to work effectively together in joint operations. In the early 1980s, the U.S. military experienced a failed joint operation in the Iranian desert to rescue hostages (Operation Eagle Claw in 1980), a terrorist bombing (Beirut in 1983) that claimed 240 servicemen, and operations in Grenada that served to highlight flaws in the conduct of joint operations. The challenges that characterized the conduct of the operations included a lack of clear lines of authority, poor command and control procedures, and incompatible communications equipment.

Prior to GNA enactment, while the unified combatant commanders (UCCs) had geographic area of responsibility (AOR) and authority, the services exercised control of the assigned component forces in a UCC's AOR. Control and movement of component forces in and out of theater are now under the direct control of the UCC. The services also had maintained their distinct cultures (and rivalries) and tradition. Rivalries were pronounced and much effort and resources were expended to sustain and protect their respective missions and capabilities. Prior to the GNA, the programming and budgeting process was a battle along service lines to support single-service missions. Reform and leadership were needed to effect change.

Reforms were needed to support more effective joint operations, including the need to establish requirements for joint officer management and joint professional military education. These reforms were encompassed as part of the Goldwater-Nichols Department of Defense Reorganization Act of 1986.

The process that led to the GNA began when General David Jones, the Chairman of the Joint Chiefs of Staff, went before the House Armed Services Committee in a closed session on February 3, 1982, about five months prior to his retirement, and said, essentially, "The system is broken. I have tried to reform it from inside, but I cannot. Congress is going to have to mandate necessary reforms." General Jones was the catalyst, the most important factor in ultimately bringing about the GNA.[1]

Congressional intent[2] in enacting the GNA was to

- Reorganize the Department of Defense (DoD) and strengthen civilian authority
- Improve military advice to the President, National Security Council, and Secretary of Defense
- Place clear responsibility on commanders of unified and specific combatant commands for the accomplishment of missions assigned to those commands
- Ensure that the authority of the commanders of unified and specified combatant commands is fully commensurate with the responsibility of those commanders for the accomplishment of missions assigned to their commands
- Increase attention to the formulation of strategy and to contingency planning
- Provide for more efficient use of defense resources
- Improve joint officer management policies

---

[1] James R. Locher III, "Has It Worked? The Goldwater-Nichols Reorganization Act," *Naval War College Review*, Vol. 54, No. 4, Autumn 2001.

[2] Goldwater-Nichols Department of Defense Reorganization Act of 1986, Public Law 99-433, October 1, 1986.

- Enhance the effectiveness of military operations and improve the management and administration of DoD.

Goldwater-Nichols and Joint Officer Management

The GNA forged a cultural revolution by improving the way DoD prepares for and executes its mission; it is the driving force behind joint officer management today. Title IV of the GNA contains the personnel-related provisions, including management policies, promotion objectives, and education and experience requirements for officers assigned to joint duty assignments (JDAs). The summarized objectives of the GNA are to

- Enhance joint warfighting capabilities
- Increase the quality of officers in joint assignments
- Ensure that officers are not disadvantaged by joint service
- Ensure that general and flag officers are well rounded in joint matters
- Enhance the stability and increase the joint experience of officers in joint assignments
- Enhance the education of officers in joint matters and strengthen the focus of professional military education in preparing officers for JDAs.

The act was a landmark document that changed the way officers are managed, and it provided specific goals that must be met. The GNA has driven changes in the way that officers are educated, trained, and experienced in joint operations, and successes have been achieved.

The intent of the joint officer management provisions was to enhance the quality, stability, and experience of officers in joint assignments, which in turn would improve the performance and effectiveness of joint operations.[3]

---

[3] Harrell et al. (1996).

Requirements of Goldwater-Nichols

The GNA established and directs numerous policies for the management of joint officers. Current statutory requirements are contained in United States Code Title 10, Chapter 38–Joint Officer Management. The following is a summary of the current joint officer management provisions:

- The Secretary of Defense shall establish policies, procedures, and practices for the effective management of officers trained in and oriented toward joint matters (joint specialty officers [JSOs]).
- The number of JSOs shall be large enough to meet the requirements for officers needed in the JDAs.
- Officers selected to be JSOs must meet education and experience requirements.
- One-half of the JDA positions above pay grade O-4 must be filled by JSOs or JSO nominees.[4]
- The Secretary of Defense shall establish career guidelines for the selection, education, training, and types of assignments for JSOs.
- Officers in pay grade O-3 may perform qualifying service in a JDA.
- Promotion objectives: Officers who served on the Joint Staff and all JSOs (after December 27, 2004) must be promoted at a rate commensurate with their service headquarters staff. Officers who have served or are serving in other JDAs must be promoted comparably to their service competitive group.
- Education: All O-7s shall attend the Capstone course; curriculum offered at joint professional military education (JPME) schools shall be periodically reviewed to strengthen joint content and prepare officers to serve in JDAs; JSOs who graduate from JPME Phase II must be assigned to a JDA; and JPME II must be three months in duration.

---

[4] JSO nominees are officers who have completed JPME II but have not completed a JDA.

- Length of JDA: two years for O-7 and above, three years for all other officers except critical occupational specialty (COS) officers, whose tour lengths are two years. Overall JDA tour length averages for each fiscal year shall be two years for O-7s and above and three years for all other officers. A full JDA tour is one that meets standard tour lengths above or through cumulative service in qualifying JDAs. Constructive credit of up to 60 days can be awarded to officers removed from a JDA for military necessity. Joint duty credit may be awarded to officers serving in qualifying joint task force (JTF) assignments.
- The Secretary of Defense shall establish procedures for overseeing the careers of JSOs and officers who serve in joint assignments and shall support the Joint Staff's capability to monitor promotions and career assignments of those officers.

For many of the statutory requirements directed in joint officer management provisions above, there are waiver authorities in Title 10 that the Secretary of Defense and/or the Chairman of the Joint Chiefs of Staff (CJCS) may apply. For example, the Secretary may waive requirement that

- a tour of duty in a JDA be performed after the officer completes JPME if it is in the interests of sound personnel management
- in the case of an officer who has completed two JDAs and is nominated for JSO, that an officer successfully completes JPME if it would be impractical or the JDAs were of sufficient breadth to prepare the officer for the joint specialty
- a JSO be assigned to a critical JDA
- newly selected O-7s attend the Capstone course
- JSOs who graduate from a JPME school be assigned to a JDA
- the length of JDAs for general and flag officers shall be not be less than two years, and for other officers shall be not less than three years.

The annual use of waiver authorities is contained in the GNA implementation report in the Secretary of Defense's Annual Report to the President and the Congress.

Reporting Requirements and Metrics of Goldwater-Nichols

To monitor the effectiveness of the statutory joint officer management guidelines, Congress requires that the Secretary of Defense submit a GNA implementation report[5] in the Annual Report to the President and the Congress. This report contains the directed joint officer data, information, and comparative data deemed necessary by the Secretary to demonstrate the performance of DoD and the services in implementing the GNA. These data are metrics by which Congress evaluates DoD's compliance with the GNA. The metrics contained in the most recent Joint Officer Management Annual Report[6] are as follows:

- summary of JSO and JSO designations, by service
- critical occupational specialties, by service
- JSOs, by branch and grade
- summary of officers on active duty with a critical occupational specialty
- summary of JSOs with critical occupational specialties[7] who are serving or have served in a second joint assignment
- analysis of the assignment where officers were reassigned on their first assignment following designation as a JSO
- average length of tour of duty in joint duty assignments
- summary of tour length exclusions

---

[5] United States Code, Armed Forces, Subtitle A, General Military Law, Title 10, Part II, Personnel, Chapter 38, Section 667, *Annual Report to Congress*.

[6] Office of the Secretary of Defense, *Annual Report to the President and the Congress*, 2003.

[7] COS officers are discussed more extensively later in this appendix.

- joint duty position distribution by service
- critical position summary
- reasons for filling critical positions with officers who are not JSOs
- list of organizations that have joint duty critical positions, which are filled by officers who do not possess the joint specialty
- comparison of waiver usage
- JPME II summary
- reasons for students not completing resident professional military education prior to attending JPME II
- temporary JTF credit
- operations for which JTF credit has been awarded
- Army, Navy, Air Force, and Marine Corps joint officer promotion comparisons.

In view of these reporting requirements, the joint officer management system can be seen as a system of inputs and outputs (officers serving in and completing JPME II and JDAs), set points and gauges (congressional guidelines and metrics), and overrides (waivers and exemptions). A complex set of rules guides the system, and the reporting requirements are complex and time consuming to evaluate and complete. However, if Congress felt the services would comply as a matter of course, reporting requirements would not be needed.

Joint Duty Assignments, the Joint Duty Assignment List, and Critical Billets

A JDA is an assignment to a billet in a multiservice or multinational command/activity that is involved in the integrated employment or support of the land, sea, and air forces of at least two of the three military departments.[8] The duties of an officer in a qualifying JDA involves producing or promulgating national military strategy, joint

---

[8] U.S. Department of Defense, DoD Instruction 1300.20, "DoD Joint Officer Management Program Procedures," December 20, 1996.

doctrine, joint policy, strategic plans or contingency plans, or commanding and controlling operations under a combatant command. Assignments to an officer's own military department or assignment for joint education or training do not qualify and are not covered by this definition. Successful completion of a JDA is a criterion for designation as a JSO.

The Joint Duty Assignment List (JDAL) is a consolidated roll that contains all the billets[9] that are approved JDAs for which joint credit can be applied. Billets are added to and deleted from the JDAL, and there is a validation process to review positions nominated for addition. A joint duty validation board, composed of representatives of Office of the Secretary of Defense, the Joint Staff, and the military departments, considers the joint content of nominated billets. A billet is evaluated and voted on according to its merit for inclusion or exclusion to the JDAL by the validation board.

Joint duty credit is the joint credit granted for the completion of an assignment (or accumulation of sufficient time in assignments) to a JDA that meets all statutory requirements.[10] There are two types of positions on the JDAL: the standard joint duty position and joint critical positions. Any qualified officer may serve in the standard JDA,[11] while fully qualified JSOs may fill the critical JDAs. These critical assignments are JDA positions in which either the incumbent should be experienced and educated in joint matters or the position would be greatly enhanced by an officer with the joint experience and education. Critical positions are proposed by heads of joint activities, approved by the Principal Deputy Under Secretary of Defense for Personnel and Readiness (PDUSD[P&R]) with the advice and assistance of the CJCS, and documented in the JDAL. Critical positions

---

[9] In the past, only 50 percent of the positions in defense agencies could qualify as JDAs, while 100 percent of the positions in other joint organizations were on the JDAL. These limits no longer exist.

[10] Chairman of the Joint Chiefs of Staff, CJCS Instruction 1331.01B, "Manpower and Personnel Actions Involving General and Flag Officers," August 29, 2003.

[11] Additionally, there are provisions for awarding of joint creditable service for duty performed in approved JTF headquarters assignments. Constraints and limitations are applicable to each period of creditable service, and waivers can be applied, with certain provisions.

are to be filled by JSOs unless waived by the CJCS[12]. There are 807 critical JDAs per the current GNA implementation report (2003).

Critical positions are not necessarily the most key and essential billets in an activity.[13] While critical billets are designated to be filled by JSOs, our discussions with Joint Staff officials indicate that JSOs are not always sought out by UCCs and other joint commands to fill critical JDAs. Other officers, with rich and current service experience, are sometimes more valued to fill these positions. For example, from 1999 to 2003,[14] of the approximately 800 critical positions on the JDAL, 241, 254, 254, 321, and 242 critical positions were filled by non-JSOs, respectively. While other reasons are reported for not filling critical positions with JSOs, 176, 184, 103, 275, and 196 critical position JSO requirements were waived, respectively, due to the "best qualified officer not being a joint specialist." Without further study, one can only speculate on the future demand for JSOs at joint commands. It has been suggested that the view expressed in unified command headquarters is that there is little distinction between the performance of JSOs and non-JSOs on the job, and little need for critical JDAs.[15]

Critical Occupation Specialty Officers

A COS is a military occupational field that involves combat operations within the services and in which the Secretary of Defense has determined that a shortage of trained officers exists. Specialties that may be designated as "COS" are determined each year. The current designated COSs for the services (per the Secretary of Defense's

---

[12] U.S. Department of Defense (1996).

[13] U.S. Department of Defense (1996).

[14] Derived from the "GNA Implementation Report" in the Secretary of Defense's Annual Report to Congress for years 2000 through 2004.

[15] Booz Allen Hamilton (2003).

Annual Report to the President and the Congress for 2004) are as follows:

- Army—infantry, armor, artillery, air defense artillery, aviation, special operations, combat engineers
- Navy—surface, submariner, aviation, SEALS, special operations
- Air Force—pilot, navigator, command/control, operations, space/missile operations
- Marine Corps—infantry, tanks/AAV, artillery, air control/air support, anti-air warfare, aviation, engineers.

The Secretary of Defense may reduce the JDA tour lengths of COS officers to two years, as long as such "COS takeouts" do not exceed 12.5 percent of the officers serving as JSOs and JSO nominees.

Joint Professional Military Education

Professional military education enhances an officer's knowledge of military science and the art of war, and there is a continuum of education that officers receive throughout their career. JPME focuses specifically on joint matters. JPME instruction that qualifies an officer for JSO/JSO nominee is performed both at the military service colleges (resident and nonresident) and at National Defense University. JPME Phase I is incorporated into the curricula of the military service colleges at both the intermediate level (O-4) and senior level (O-5 and O-6). The Joint Forces Staff College (JFSC) provides JPME Phase II[16] to both intermediate- and senior-level students.

Intermediate-level colleges teach joint operations from the standpoint of service forces in a joint force supported by service component commands. Senior-level service colleges address theater and

---

[16] The JPME II course of instruction at JFSC was recently reduced in duration from 12 weeks to 10 weeks, which allows for an additional session to be held each year. Liaison with JFSC officials indicates that four sessions are now held (beginning in FY 2005), with the maximum capability of 255 students per session.

national-level strategies and processes. Curricula focus on how unified combatant commanders, Joint Staff, and DoD use the instruments of national power to develop and carry out the National Military Strategy.[17]

The Joint and Combined Warfighting School at the JFSC (for JPME II credit) provides instruction in joint operations from the perspective of the CJCS, UCCs, and JTF commanders. The course develops joint attitudes and perspectives and exposes officers to other service cultures while maintaining a concentration on joint staff operations.

The National War College (NWC) and Industrial College of the Armed Forces (ICAF) course of instruction provides full JPME credit for graduates.

To meet the educational prerequisites to become a JSO/JSO nominee, officers must complete one of the following:

- JPME I at an accredited service intermediate- or senior-level college, followed by JPME II at JFSC[18]
- an intermediate- or senior-level international military education program for which JPME I credit has been approved by the CJCS, followed by JPME II at JFSC
- the course of instruction at either NWC or ICAF.

Joint Specialty Officer Designation

While there is no longer a board process for JSO selection, officers who complete JPME II and JDAs are nominated to the Office of the Secretary of Defense for JSO designation. There are four categories of officers who are considered for JSOs:

---

[17] Chairman of the Joint Chiefs of Staff, CJCS Instruction 1800.01B, "Officer Professional Military Education Policy," August 30, 2004.

[18] Attendance at JPME II prior to completing JPME I requires a waiver by CJCS.

- officers who complete JPME I and II and a full JDA (36 months), or COS officers who complete a 36-month JDA before completing JPME I and II
- COS officers who complete JPME I and II and a 22–24 month tour[19]
- other officers who complete a JDA prior to JPME II (requires a waiver)
- officers who complete two full JDAs but no JPME II.[20]

The total number of waivers granted for officers in last two categories for a fiscal year may not be greater than 10 percent of the total number of officers in that pay grade selected for the joint specialty during that fiscal year.

Cumulative Joint Duty Credit

The 1996 NDAA authorized that credit for a full JDA or credit countable for determining cumulative service[21] is awarded to officers serving in qualifying temporary JTF assignments. Cumulative credit may be earned in one of two methods:

- For service performed in a JDA that totals less than two years (general and flag officers) and three years (other officers) and includes at least one tour of duty in a joint assignment that was

---

[19] COS officers can receive up to two months constructive credit if they depart their JDA for "military necessity."

[20] The Secretary of Defense may waive that an officer successfully complete a program of education if impractical to do so at the stage of his or her career and the joint duty performed was of sufficient breadth to prepare the officer for the joint specialty.

[21] JDA credit awarded for Certain Task Force assignments is exempt from JDA promotion reports, minimum tour length requirements and military service tour length averages, assignment fill rates, and professional joint education sequencing requirements (U.S. Department of Defense, 1996).

either performed outside the continental United States or terminated because of a reassignment[22]
- For service in combined joint task force (CJTF) headquarters assignments in approved operations.

Cumulative joint duty credit for continuous service performed in a JDA is granted for tour lengths of at least 10 months but less than the time needed to qualify for full joint duty credit. The Secretary of Defense retains waiver authority for tour lengths of JDAs.

The award of cumulative credit for JTF headquarters duty in ongoing operations is currently authorized for positions in CJTF-AFG (Afghanistan), CJTF-HOA (Horn of Africa), and CJTF-7 (Iraq), and other JTFs. Restrictions pertain to the award of CJTF cumulative credit, specifically:

- The minimum JTF service time is 90 days.
- The single cumulative credit for duty performed in these operations does not satisfy the requirement for eligibility for promotion to O-7, selection as a JSO, or inclusion in joint promotion reports.
- 36 months of cumulative credit is needed for full JDA credit (general and flag officers and COS officers can obtain full credit after earning 24 months of cumulative credit[23]).
- Credit is limited to officers assigned to JTF headquarters staff, not to subordinate organizations or service components.
- Officers serving in grade O-3 who are filling an O-4 or above billet qualify for joint duty credit. Officers in grades O-1 and O-2 or any officer filling an O-3 or below billet does not qualify for credit.

---

[22] A reassignment must be either for a personal reason beyond the officer's control or if an officer was promoted to a grade in which a JDA was not available to him within a command at his promoted grade or the billet was eliminated.

[23] Joint Staff, "Cumulative Joint Duty Credit for Combined Joint Task Force Headquarters Assignments in Approved Operations," memorandum, J-1A 00165-04, July 12, 2004.

- Only active duty officers are eligible. Reserve component and "professional" specialties in the medical, legal, and religious career fields are excluded.
- Only one officer per qualifying billet may be awarded credit.
- Award of credit is not automatic. Officers serving in JTF headquarters prior to July 1, 2004 must apply for credit.

Awarding of Joint Task Force Duty Credit Is an Issue Today and for the Future

As discussed in Chapter Two, joint task forces are the organizational response to military missions. Many current issues related to joint officer management center around the awarding of joint duty credit for officers who serve either in JTF headquarters or in service component commands and service units assigned to JTF. This is an issue of great concern to senior leaders and officers who perceive that they experience jointness but do not receive credit for it during on-going operations such as in Afghanistan and Iraq. Focusing on the administrative process of "credit" rather than actual joint experience received appears misdirected, but it is a reflection of the current system, which does not always adequately acknowledge the validity of joint experience when gained with a JTF.

Officers are increasingly likely to have assignments in JTFs (headquarters and service components), and officers performing duty in these assignments gain valuable joint experience. The value of the experience gained by an individual officer in a JTF assignment will enhance that officer's involvement and support in future joint operations. Awarding joint service credit provides recognition of an officer's joint experience in an operational billet responsible for planning and conducting joint operations. While the joint experience in some JDAs may not have as great of a joint content, the planning and melding of joint capabilities by a JTF staff in task force operations is the direct application of the objectives behind the enactment of the GNA. JTF staff plan, coordinate, and direct the combined capabilities of the task force to achieve the combatant commander's intent.

These high-intensity joint operations provide an officer with a rich joint experience that cumulative credit does not adequately capture.

During interviews with combatant command and service officials, a recurring theme among them was that "everything we do is joint." The challenge is that, while temporary JTF headquarters assignments may provide an officer with a rich joint experience, the credit for the content of the experience is not appropriately recognized. Determining what positions provide valid joint experience and how much joint experience is necessary for which positions is the crux of a strategic approach.

Collecting Data on Joint Experience

Our study is aimed at a one-time gathering of data on billets that require joint experience, education, or training and billets that provide such experience. Eventually, such data need to become routine and standardized within the military officer classification systems. Each service has a means for collecting data elements about billet requirements and officer competencies; these systems are reviewed below. Each is different from the other in terms of operation and data elements collected. The long-term solution is the Defense Integrated Military Human Resources System (DIMHRS) that will fully track all skill sets: those competencies required by the position and those held by the service member.[1]

Navy

The Navy uses additional qualification designation (AQD) codes to identify more specifically the qualifications required by a billet and the qualifications of officers. These three-digit AQD codes can be

---

[1] "Combatant commanders and other DoD managers very often require specific skill sets for mission-essential operations. Multiple personnel systems provide inconsistent data of variable accuracy across the services and the managers are dependent on the individual services to search multiple databases to identify qualified individuals" (U.S. Department of Defense, "Revolutionizing Military Personnel and Pay," DIMHRS fact sheet, Office of the Under Secretary of Defense for Personnel and Readiness, undated).

awarded to the incumbent of a designated billet, or an officer may qualify through education, training, and experience.

Officer AQD

> When entered in an officer's record, the AQD code identifies the attainment of skills and knowledge as recognized by competent authority, in addition to those identified by the officer designator, grade, NOBC or subspecialty.[2]

Billet AQD

> When applied to a billet on a manpower authorization, the AQD code generally identifies a requirement for skills and knowledge needed to perform the duties and/or functions of a billet additional to those identified by the billet designator, grade, NOBC or subspecialty. The AQD generally indicates a requirement for an officer who has attained special qualifications through training and/or experience.[3]

There are provisions for establishing new AQD. One criterion is that an AQD must have a practical application in planning, personnel control, career management, training, experience, or manpower information functions.

Joint Specialty AQD

Four two-letter AQD codes with the first character "J" are used for the Navy's joint duty/joint specialty field. "JD" is used to designate joint billets external to the service but is not used not for officers. "JJ" is used for both billets and officers to designate Joint Operation Planning and Execution System (JOPES) proficiency. "JP" is used for both billets and officers to designate operational planning expertise. "JS" applies only to officers and designates their joint experience or

---

[2] Department of the Navy, *Manual of Navy Officer Manpower and Personnel Classifications*, NAVPERS 15839I, October 2003a.

[3] Department of the Navy (2003a).

education. For each of these two-letter codes, a third digit can further refine the information.

The JJ code provides an example of use. Office of the Chief of Naval Operations (OPNAV) Instruction 1521.2 (1995) identifies common, minimum requirements to certify individuals as JOPES experts.[4] Unified combatant commanders and naval component commanders identified billets that provide naval personnel with sufficient on-the-job training for them to achieve JOPES qualification. The instruction provides a list of such billets (both inside and external to the Navy). In addition, the UCCs and NCCs identified the minimum requirements for JOPES qualification, and these are also listed. Officers who complete tours in the designated billets and officers whose commands certify that they meet the JOPES personnel qualification standard are awarded the AQD. Code JJ1 is used by the UCCs and NCCs to identify billets requiring the competency. As stated above, officers who serve in billets listed in OPNAV Instruction 1521.2, or who qualify through other education and experience, receive the AQD. We used a Defense Manpower Data Center (DMDC) data system to identify three Navy commands (COM-PACFLT, COMUSNAVSO, and CUSNFS Det Miami) that have identified a total of 10 billets requiring the expertise.

In essence, use of the JJ AQD is a microcosm of how a routine and standardized joint management system might operate. Billets both inside and external to the Navy requiring the competency are determined by commands requiring the competency. Completion of a tour in predesignated qualifying billets inside and external to the service provides the officer with the qualification. Alternative means for qualifying through education/training and demonstrated expertise (e.g., performing duties on a JTF staff) are provided.

---

[4] See U.S. Department of Defense, "Audit of the Joint Operation Planning and Execution System," Officer of the Inspector General, 94-160, June 30, 1994, which tasked each service to identify and track individuals with JOPES expertise.

Summary

To us, it appears the Navy officer classification system could most easily accommodate incorporating additional information about billets and officers. The use of additional third characters can refine the data elements sufficiently, and a process is in place for designating either billets or officers, or both.

Marine Corps

The Marine Corps uses a framework of occupational fields and military occupational specialties (MOSs) to identify the skills of individuals and the requirements of organizations.[5] Four-digit codes identify Marine Corps personnel duties, skill attributes, and requirements within specific functional areas. Occupational fields are broad categories of skills and make up the first two digits of the MOS. The MOS describes a group of skills and related duties and consists of a four-digit code and a descriptive title.

Other MOS Categories

There are two specialized uses of MOSs: skill designator MOSs and identifying/reporting MOSs. Skill designator MOSs are not primary MOSs but are usually those specialties with low density of either requirements or officers. For example, the JOPES skill is designated through a skill designator MOS. MOS 0502 (force deployment planning and execution officer) is assigned to officers and is also used to identify billets in organizations requiring special planning skills and experience for JOPES. There are requirements and prerequisites that an officer must meet before being assigned the MOS; however, the MOS would not be considered a primary MOS. In the DMDC database, 52 billets are shown as requiring MOS 0502.

Identifying/reporting MOSs are used to identify skills of individuals and to identify billets in tables of organization. MOS 9701

---

[5] Department of the Navy, *Military Occupational Specialties Manual*, Marine Corps Order P1200.7Y, April 2003b.

(joint specialty officer nominee) and MOS 9702 (joint specialty offi-cer) are two of these MOSs. These MOSs are never to appear as a primary or additional MOSs and are used primarily to identify offi-cers for assignment purposes.

Summary

The Marine Corps system is not as refined as that of the Navy but could accommodate more routine designation of billets and officers with specific categories of joint experience, education, and training. For example, additional skill designator MOSs could be used for both billets with particular needs and officers with certain qualifications.

Army

The Army classification system uses branches, areas of concentration, and skill identifiers for positions requiring officer skills and officer personnel based on qualifications.[6] The first three digits of the MOS designate area of concentration (of which the first two digits desig-nate branch). Skill identifiers are separate two-character codes used to identify the skills required of a position as well as the skills in which officers may be classified. Skill identifiers identify specialized occupa-tional areas that are not normally related to any one particular branch of area of concentration but are required to perform the duties of a special position. Skill identifiers may require significant education, training, or experience.

The Army uses skill identifiers to categorize both billets and officers for JDA qualified (3A); joint planner (3H); joint command, control, and communications (3K); and JSO (3L). Detailed descrip-tion of positions and officer qualifications are provided. Currently 3A and 3S are only used for positions on the JDAL, with 3S designating critical positions. Codes 3H and 3K can be used for any Army billet

---

[6] Department of the Army, Army Regulation 611-1, "Military Occupational Classification Structure Development and Implementation," September 1997; Department of the Army, "Military Occupational Classification and Structure," Army Pamphlet 611-21, March 1999.

either inside or external to the service. The Army has designated about 4,800 billets (O-3 through O-10) and about 3,300 officers with these codes, as shown in Table B.1.

Summary

It is likely that the Army could modify its system of skill identifiers to accommodate more nuanced use of joint experience, education, and training by either modifying existing skill identifiers or adding new ones.

Table B.1
Use of Skill Identifiers

Billets					

Army active duty officer positions (billets) in grades O-3 and higher that have additional skill identifiers of 3A, 3H, 3K, and 3L

Grade	Additional Skill Identifier				Total (#)
	3A (#)	3H (#)	3K (#)	3L (#)	
O-3	1	195	76	0	272
O-4	1,357	327	185	3	1,872
O-5	1,465	161	98	166	1,890
O-6	489	59	29	184	761
O-7	14	0	3	1	18
O-8	10	0	0	2	12
O-9	3	0	0	2	5
O-10	4	0	0	0	4
Total (#)	3,343	742	391	358	4,834

Personnel					

Army active duty officer personnel in grades O-3 and higher who have additional skill identifiers of 3A, 3H, 3K, and 3L

Grade	Additional Skill Identifier				Total (#)
	3A (#)	3H (#)	3K (#)	3L (#)	
O-3	1	3	2	0	6
O-4	179	369	4	0	552
O-5	1,403	81	10	3	1,497
O-6	1,033	18	6	12	1,069
O-7	69	0	1	17	87
O-8	55	4	1	15	75
O-9	23	4	0	4	31
O-10	9	0	0	0	9
Total (#)	2,772	479	24	51	3,326

Air Force

The Air Force officer classification system is specified in greater detail and is more complex than the other services' but provides less centralized information about jointness. In the Air Force classification system, there is nothing comparable to the AQD of "J" in the Navy, the ASI of "3" in the Army, or the MOS of 9702 in the Marines.

An air force specialty (AFS) is a grouping of positions that require similar skills and qualifications.[7] These specialties are further grouped into utilization fields and career areas. For example, the first digit of the four-digit AFS code (AFSC) identifies the career group such as operations, logistics, or support; the second digit identifies the utilization field such as pilot in the operations career group; the third digit identifies a functional area such as airlift; and the fourth digit identifies the qualification level such as entry or intermediate. The AFSC is the backbone of the classification system that also includes prefixes, suffixes, special duty identifiers, reporting identifiers, and special experience identifiers. The prefix identifies an ability, skill, or special qualification not restricted to a single AFSC. These prefixes can be used to identify position on manning documents and officers qualified to serve in such positions. A suffix (also called a "shredout") identifies positions associated with particular equipment or functions within a single specialty. Special duty identifiers classify duties and responsibilities not clearly within a specific career area; reporting identifiers classify people or positions that cannot be identified elsewhere in the classification system; special experience identifiers classify special experience and training not otherwise identifiable in the classification system.

Identifying Jointness

No one code identifies either positions or officers associated with joint experience or education. The closest is the prefix "R," which is

---

[7] Information and quotes in this section are from Department of the Air Force, Air Force Instruction 36-2101, "Classifying Military Personnel," April 2001, and Department of the Air Force, *Officer Classification,* Air Force Manual 36-2105, April 2003.

used for positions and officers for contingency and war planners, including but not limited to JOPES. Officers can qualify by completing a course of training and serving 6 months in an R position or by serving 12 months in the R position. Some prefixes have restrictions on their use that might be useful if a prefix were identified for joint experience. For example, the use of a prefix on manpower documents could be restricted solely to authorizations in unified commands, headquarters Air Force, major command centers, numbered air forces, joint staff, etc.

Special experience identifiers (SEIs) would seem to be a likely way to incorporate information in the Air Force system about jointness required or provided. For example, the following example is given on reviewing positions to determine appropriateness of identifying positions by SEI: Is the officer filling the position gaining special experience identified by a current SEI? Does the position require special experience identified by a current SEI? If the answer to either of these questions is yes, then coding the position with the SEI is appropriate. The three-digit SEI code consists of a one-digit activity code and a two-digit experience set. The activity code identifies possible activities performed in a variety of utilization fields, and the experience set identifies a particular system, level of experience, or the type of experience. For example, EJ is an experience set for JOPES that could be combined with an activity such as operations or tactical analysis. More than 46,000 separate SEI codes are possible using already defined elements. Moreover, the Air Force cautions that there are no formal programs established to use SEIs in the officer assignment process: "The size of such a system, coupled with no edits on officer SEIs to any particular AFSC, limits ability of any single agency to control placement of SEIs on unit manpower documents." In essence, the SEI, where it exists, is simply used to embellish the other data fields in the classification structure.

How then does the Air Force identify billets and officers? Rather than making it an explicit part of the officer classification system for billets and officers, the Air Force uses data fields in its personnel data system to designate particular characteristics dealing with an officer's assignment history. Thus, one can observe about Air Force officers the broad occupational specialty in which they served their joint tour (e.g., technical, intelligence, communications), whether it was a CONUS or overseas tour, the tour credit status, the reason for early departure, and JSO status.

Summary

This information is not dissimilar from what is available in the other services. There is simply not as straightforward a way to designate billets that require or officers who have had a joint experience and to use that information contemporaneously. For example, in assigning officers, they should be matched to a position number that requires their AFSC. But nothing in the AFSC for either the position or offi- cer would currently allow a match of joint experience with a need for it. Information about the billet need would have to be communicated in the requisition (e.g., in major command remarks) each time an officer was needed. And the assignment officer would have to work harder to discern in an officer's background (e.g., in an SEI if these identifiers were recorded) the prior experience or education. SEIs or prefixes might be used by the Air Force to characterize both positions and officers for joint experience, education, and training, but there may be better ways.

Data Requests to Determine Billets That Require Prerequisite Jointness and Billets That Provide Jointness

This appendix includes drafts of memos that could be distributed by the research sponsor to request the necessary data regarding positions that require jointness as well as to elicit nominations of positions not currently on the JDAL but that might provide joint experience to officers. There are two versions of the memo included here. The first version is designed for distribution to the services, inquiring which of their positions might require previous joint experience, education, or training and which of their in-service positions should be nominated for inclusion in the survey to determine which positions provide valid joint experience. The second version is designed for distribution to external organizations and requests that these organizations identify their positions that might require previous joint experience, education, and training. The second version also requests the organizations to assist in the survey of all their organization's positions in pay grades O-1 through O-6.

Draft Memo to Services

Subject: Joint Officer Management Strategic Approach

The Secretary of Defense and the Chairman of the Joint Chiefs of Staff are evaluating changes to enhance the development of the officer corps for joint matters. The Under Secretary for Personnel and

Readiness (USD[P&R]) and the Joint Staff are collaborating to develop the strategic plan for joint officer management and joint professional military education, and the [NAME OF ORGANIZATION] is conducting a study that will provide recommendations to aid in developing a strategic approach.

The missions, goals, and desired organizational outcomes through the conduct of joint operations have been recognized and incorporated into strategic plans, vision, and mission documents from the National Security Strategy and National Military Strategy through the planning guidance of the services. The National Security Strategy calls for the strengthening of joint operations; the National Military Strategy calls for U.S. armed forces to be multi-mission capable, interoperable among all elements of U.S. services and selected foreign militaries, and able to coordinate operations with other agencies of government and some civil institutions. Joint Visions 2010 and 2020 state that the joint force remains key to operational success in the future and that to be effective the force must be "intellectually, operationally, organizationally, doctrinally, and technically joint." The Secretary of Defense has identified that one of his top priorities is to strengthen combined/joint warfighting capabilities and to bring jointness to the lowest appropriate level. A common theme is that officers must be educated, trained, and experienced in joint matters to enhance the joint warfighting capability of the United States.

The development of a strategic approach to joint officer management requires that the services define their critical joint workforce characteristics needed now and in the future to achieve missions and goals and to have continued operational success in future joint operations. Such a strategic approach requires a fresh look in identifying billets that *require* some prior joint experience, education, and/or training as well as those billets that *provide* a joint experience. This strategic approach to joint officer management will evaluate and analyze billets in service organizations as well as those external to the service. This strategic approach recognizes that it is likely that some billets within service organizations may require previous joint experience and that other billets resident in service organizations could provide a joint experience, despite the current restrictions that state that only JDAL billets can provide joint experience.

We are using two separate processes to evaluate and identify billets that

a. *Require* prior joint education, experience, and/or training, or
b. *Provide* joint experience.

The following attachments provide information and procedures to follow for each process.

Memo Attachment A

The following guidance is provided to assist in determining billets that *require* prior joint education, experience, and/or training.

The study requires a collection of data and information essential for the development of the strategic approach. To implement a strategic approach, [NAME OF ORGANIZATION] needs data about which billets (including those in service organizations as well as those external to the services) *require* that the officers filling those positions have some prerequisite degree of joint experience, education, and/or training or would benefit from such prerequisites. A billet that requires joint experience or joint education implies that a prerequisite joint education and experience tour would better qualify an officer to perform the mission requirements of the position.[1] For example, it is likely that a joint experience may be important to the performance of an army brigade commander. Joint experience would likely enhance the officer's awareness of how best to employ his or her forces in joint warfighting in concert with the other services.

Enclosure to Attachment A is a listing of selected officer authorizations for lieutenant colonel/commander and colonels/navy captains. This list represents billets that the research team evaluated as in-service positions that may potentially require prior joint experience, joint education, or joint training to perform best the duties of

---

[1] Joint education refers to JPME II–level education.

the position in a joint environment. Also enclosed is a description of the analytical steps used to decrease the number of billets in the listing. Deleted billets were judged as less likely to require or benefit from prior joint experience, joint education, or joint training. However, services may include any deleted billet if they judge such billets as requiring or benefiting from prior joint experience, joint education, or joint training.

The manpower authorizations in the enclosure were arrived at by deselecting the most unlikely billets requiring joint education and experience and presenting the most likely billets for consideration by service experts. The numbers of billets were left intentionally broad for service determination and discretion in determining the requirement of jointness. If deselected billets are deemed as demanding or benefiting from officers with prior joint education and experience, request that they be added to this list. Conversely, if billets contained in the listing are deemed as not demanding or not benefiting from joint education and experience by subject-matter experts, request they be deleted or lined out from this list. This step will identify where (in what positions) officers with joint experience, education, and training are needed and the number of positions needed. Please note that [NAME OF ORGANIZATION] does not need precise identification of all like billets for its analysis. For example, if 50 percent of battalion commanders require joint experience, education, and/or training, then specifying the number of billets by grade, occupation, and type position (e.g., battalion command) will provide sufficient data for analysis.

The goal of the data collection is to understand the need for prior joint experience, joint education, or joint training to perform best the current and future mission requirements of the billet. Request the services review the enclosure and identify billets that require prior joint education, joint experience, and/or joint training. For each billet identified as requiring or benefiting from prior joint education, joint experience, and/or joint training, indicate whether the following kinds of experiences are required, desired, or neither required nor desired:

a. multiservice experience (MS)
b. interagency experience (IA)
c. multinational experience (MN)
d. job-specific joint training (JT) (such as that concerning specific systems or processes used)
e. general joint education (JE).

Complete the appropriate columns next to each billet with an "R" as required and "D" as desired, and leave blank to indicate as neither required nor desired.

Identification of these elements is important in order to address proposals either to limit the definition of what constitutes jointness and/or readdress the education and training provided for joint officers.

Memo Attachment B

The following guidance is provided to assist in determining which billets *provide* a joint experience.

Officers serving in positions on the Joint Duty Assignment List (JDAL) presently receive joint credit. One frequent criticism of the current system for awarding joint duty credit is that officers serve in other assignments not on the JDAL that provide a rich joint experience but which do not grant the officer joint duty credit. Likewise, there are officers serving in assignments on the JDAL that may not provide a rich joint experience, either because of the content of the work or limited interaction with the other services, nations, or agencies.

The current definition of joint assignments states that a billet that provides a joint experience can be thought of as assignment to an activity or multinational command that is involved in the integrated employment or support of the land, sea, and air forces of at least two of the three military departments. The preponderance of the officer's duties involves producing or promulgating the National Military Strategy, joint doctrine, joint policy, strategic plans, or contingency plans, or to commanding and controlling operations under a combatant command.[2] Because this research will explore different definitions of jointness, we encourage the services to think inclusively about billets that provide their incumbent with multinational, multiservice, or interagency experience, at any officer pay grade up to and including O-6.

The objective of this data collection and subsequent analysis is to identify positions that provide officers with significant experience in joint matters and thus should be deemed joint duty assignments. This data gathering and analysis will consider positions in service organizations and external organizations. The services will nominate candidate in-service billets that likely provide a joint experience. External positions, including those currently on the JDAL as well as non-JDAL billets in external organizations, will be surveyed through each external organization.

Request that [NAME OF ORGANIZATION] be provided with a list of in-service candidate billets, in Excel format, that services consider provide a valid joint experience. Request billet information to include organization, billet number, grade, occupation, and [TO BE DETERMINED]. These elements are important for addressing proposals to limit the definition of what constitutes jointness and/or readdress the education and training provided for joint officers.

Additionally, we will require the service to instruct each of the incumbents of specified billets to complete a Web survey during October 2004, so the service will need to be able to contact these incumbents individually.

---

[2] Derived from the definition of a Joint Duty Assignment contained in U.S. Department of Defense (1996).

Draft Memo to External Organizations

Subject: Joint Officer Management Strategic Approach

The Secretary of Defense and the Chairman of the Joint Chiefs of Staff are evaluating changes to enhance the development of the officer corps for joint matters. The Under Secretary for Personnel and Readiness (USD[P&R]) and the Joint Staff are collaborating to develop the strategic plan for joint officer management and joint professional military education, and the [NAME OF ORGANIZATION] is conducting a study that will provide recommendations to aid in developing a strategic approach.

The missions, goals, and desired organizational outcomes through the conduct of joint operations have been recognized and incorporated into strategic plans, vision, and mission documents from the National Security Strategy and National Military Strategy through the planning guidance of the services. The National Security Strategy calls for the strengthening of joint operations; the National Military Strategy calls for U.S. armed forces to be multi-mission capable, interoperable among all elements of U.S. services and selected foreign militaries, and able to coordinate operations with other agencies of government and some civil institutions. Joint Visions 2010 and 2020 state that the joint force remains key to operational success in the future and that to be effective the force must be "intellectually, operationally, organizationally, doctrinally, and technically joint." The Secretary of Defense has identified that one of his top priorities is to strengthen combined/joint warfighting capabilities and to bring jointness to the lowest appropriate level. A common theme is that officers must be educated, trained, and experienced in joint matters to enhance the joint warfighting capability of the United States.

The development of a strategic approach to joint officer management requires that the services define their critical joint workforce characteristics needed now and in the future to achieve missions and goals and to have continued operational success in future joint operations. Such a strategic approach requires a fresh look in identifying billets that *require* some prior joint experience, education, and/or training as well as those billets that *provide* a joint experience. This strategic approach to joint officer management will evaluate and analyze billets in service organizations as well as those external to the service. This strategic approach recognizes that it is likely that some billets within service organizations may require previous joint experience and that other billets resident in service organizations could provide a joint experience, despite the current restrictions that state that only JDAL billets can provide joint experience.

We are using two separate processes to evaluate and identify billets that

a. *Require* prior joint education, experience, and/or training, or
b. *Provide* joint experience.

The following attachments provide information and procedures to follow for each process.

Memo Attachment A

The following guidance is provided to assist in determining billets that *require* prior joint education, experience, and/or training.

The study requires a collection of data and information essential for the development of the strategic approach. To implement a strategic approach, [NAME OF ORGANIZATION] needs data about which billets (including those in service organizations as well as those external to the services) *require* that the officers filling those positions have some prerequisite degree of joint experience, education, and/or training or would benefit from such prerequisites. A billet that requires joint experience or joint education implies that a prerequisite joint education and experience tour would better qualify an officer to perform the mission requirements of the position.[3] For example, it is likely that a joint experience may be important to the performance of an army brigade commander. Joint experience would likely enhance

---

[3] Joint education refers to JPME II–level education.

the officer's awareness of how best to employ his or her forces in joint warfighting in concert with the other services. Likewise, there may be billets in joint organizations that are not currently labeled "critical" that require prerequisite joint experience or joint education, while the content of some "critical" billets may not have such a requirement, absent Goldwater-Nichols requirements.

Enclosure to Attachment A is a sample listing of officer authorizations for lieutenant colonel/commander and colonels/navy captains. Request that you use the format indicated in Attachment A to identify those billets in your organization requiring or benefiting from previous joint experience, joint education, and/or joint training. Request that you consider the content of each job rather than any current legal requirement (such as current designation as a "critical" billet).

For each billet identified as requiring or benefiting from prior joint education, joint experience, and/or joint training, indicate which of the following kinds of experiences are required or desired:

a. multiservice experience (MS)
b. interagency experience (IA)
c. multinational experience (MN)
d. job-specific joint training (JT) (such as that concerning specific systems or processes used)
e. general joint education (JE).

Request note the appropriate columns under each billet with an "R" as required and "D" as desired, and leave blank to indicate as neither required nor desired.

Identification of these elements is important in order to address proposals either to limit the definition of what constitutes jointness and/or readdress the education and training provided for joint officers.

Memo Attachment B
The following guidance is provided to assist in determining which billets *provide* a joint experience.

Officers serving in positions on the Joint Duty Assignment List (JDAL) presently receive joint credit. One frequent criticism of the current system for awarding joint duty credit is that officers serve in other assignments not on the JDAL that provide a rich joint experience but which do not grant the officer joint duty credit. Likewise, there are officers serving in assignments on the JDAL that may not provide a rich joint experience, either because of the content of the work or limited interaction with the other services, nations or agencies.

The current definition of joint assignments states that a billet that provides a joint experience can be thought of as assignment to an activity or multinational command that is involved in the integrated employment or support of the land, sea, and air forces of at least two of the three military departments. The preponderance of the officer's duties involves producing or promulgating the National Military Strategy, joint doctrine, joint policy, strategic plans, or contingency plans, or to commanding and controlling operations under a combatant command.[4] Because this research will explore different definitions of jointness, we encourage participating organizations to think inclusively about billets that provide their incumbent with multinational, multiservice, or interagency experience, at any officer pay grade up to and including O-6.

The objective of this data collection and subsequent analysis is to identify positions that provide officers with significant experience in joint matters and thus should be deemed joint duty assignments. This data gathering and analysis will consider positions in service organizations and external organizations. The services will nominate candidate in-service billets that likely provide a joint experience. External positions, including those currently on the JDAL as well as non-JDAL billets in external organizations, will be surveyed through each external organization.

---

[4] Derived from the definition of a Joint Duty Assignment contained in U.S. Department of Defense (1996).

Thus, request that you cooperate with [NAME OF ORGANI-ZATION] as it surveys the incumbents of all officer billets, pay grades O-1 through O-6, in your organization.

For any officer billets not currently on the JDAL, request that you provide billet information to include organization, billet number, grade, occupation, and [TO BE DETERMINED]. These elements are important for addressing proposals to limit the definition of what constitutes jointness and/or readdress the education and training provided for joint officers.

RAND will provide access codes for all officers in your organization, by billet, for them to participate in a Web survey. Request you identify points of contact to coordinate survey with RAND, distribute access numbers to individuals in your organization, and confirm individual survey completion.

Draft Memo of Instruction to Survey Points of Contact

The following is a draft memo for distribution to the official coordinators or points of contact (POCs) in external organizations participating in the survey research. A similar memo would be required for distribution to the services, to the extent that they nominated billets for inclusion in the survey that will determine whether or not the billets provide a valid joint experience.

Memo for Survey Points of Contact (External Organizations)

This memorandum contains instructions for you to facilitate the survey of billets in your organization.

Please find the attached list of the billets to be surveyed in your organization. This list contains billet name, position code, password, and, in some instances, the billet incumbent (as of [DATE]). We acknowledge that this individual may no longer be the billet incumbent, but have included this name to assist you in identifying the individual billets. The information contained in the list is a compilation of data from Joint Duty Assignment Management Information System (JDAMIS) (for those billets currently on the JDAL) and information provided by your organization. We have sorted this data to print by directorate to ease your distribution of the materials.

Our intent is to, wherever possible, survey the incumbents of the billets listed. Please inform all billet incumbents to log onto the survey Web site at [WEB SITE] and complete the survey. They will need their billet code and password as listed in the attachment.

Unoccupied billets or billets with incumbents of less than two months service will need to be evaluated by another person knowledgeable about the billet. This might be a direct supervisor or a close coworker. We will rely on you, the survey POC for your organization, to identify such a person or ensure that the appropriate directorate identifies such a person. When someone other than the incumbent evaluates a billet, he will require a third identifier in addition to position code and position password. We have labeled this third identifier the "nonincumbent code." If he is also evaluating his own billet, he may use his own position code as the nonincumbent code when he evaluates other billets. If he is not evaluating his own billet (e.g., if he is a civilian), then please provide him a nonincumbent code from the list we have enclosed with this memorandum. When nonincumbents evaluate multiple billets other than their own, they use the same nonincumbent code each time. Please assign only one nonincumbent code to each individual who is evaluating other billets, regardless of how many billets he is evaluating.

Table 1
Sample Position List

JOB TITLE	NAME	POSITION CODE	POSITION PASSWORD
CH, JT DOCTRINE	Smith, Robert	AB200	20KJL
POLICY OFFICER	Wilson, James	OB500	33UPZ
POLICY OFFICER	Johnson, Mary	OB600	42TYN

Table 2
Sample Nonincumbent
Code List

NONINCUMBENT CODES
NIC123
NIC124
NIC125

For example, referring to the sample list, if Colonel Robert Smith evaluates his own billet, he would use his position code AB200 and the position password 20KJL to complete the survey. After evaluating his own billet, if Colonel Smith is also to evaluate billets OB500 and OB600, then when evaluating the first of these billets, he would enter OB500 as the position code and 33UPZ as the position password. The survey will ask him whether he is the incumbent. When he says that he is not, he would be asked whether he has (or will) also evaluated his own billet. After responding positively, Colonel Smith will then be queried for his own billet's position code and he would enter AB200. When evaluating OB600, he would enter OB600 as the billet code, 42TYN as the password, and then, when queried, his own position code of AB200. In these latter two examples (for billets OB500 and OB600), Colonel Smith is using his own position code as a "nonincumbent code" and will do so for all the billets he evaluates as a nonincumbent.

There may be instances in which a billet needs to be evaluated by an individual who is not also evaluating his own billet. For example, a DoD civilian may know the most about the billet to be evaluated. In this instance, Mr. Miller is evaluating OB500 and OB600. He logs into the Web survey and enters OB500 and the password 33UPZ. The survey inquires whether he has evaluated or will also evaluate his own billet. When he says no, his billet is not being evaluated, he will be referred to the J1 survey POC for a nonincumbent code. The survey POC in J1 should provide a nonincumbent code (e.g., NIC123) to Mr. Miller from the list provided by the RAND

Corporation and then take note not to provide that nonincumbent code to any other individuals. Mr. Miller will use that nonincumbent code for all the billets he evaluates.

We expect your organization to complete the surveys between the dates of [DATE] and [DATE]. After the start date, we will be in contact with you regarding which billets have not yet been assessed. Please refer any further questions to [EMAIL ADDRESS] or [PHONE NUMBER].

Survey Protocol

There are two versions of the survey protocol. The first version included below is the protocol version that would be completed by an individual evaluating his or her own billet. The second protocol version is that designed for individuals evaluating billets other than their own, such as in the instance of an unoccupied billet. When an individual logs onto the Web site and answers the initial survey questions, the Web site will automatically issue the correct version of the protocol, based on his or her answers to questions one through three.

Although not included in this documentation, the Web site should also include sufficient links to definitions and explanations to ease completion of the survey. This documentation includes occasional notation for the programmer, to provide insights regarding the best survey layout format and to clarify differences between the two protocol versions.

Protocol Version for Position Incumbent

1. Please enter the position number and password below:
   Position number _____
   Password _____

NOTE TO PROGRAMMER: CONFIRM AND DISPLAY BILLET DESCRIPTION.

2. Is this the billet you have been asked to evaluate?
❏ Yes
❏ No → END SURVEY. Please speak to your survey point of contact for further instructions.

3. Are you the
❏ Person occupying the billet? → Please ensure that you answer all of the following questions based on the confirmed billet, to include any TDY experiences during your service in the billet.
❏ Supervisor of the billet?
❏ Another person designated to complete the survey?

NOTE TO PROGRAMMER: IF PERSON OCCUPYING BILLET, CONTINUE WITH THIS PROTOCOL VERSION.

4. How many months have you been assigned to this billet?
_____ → *If less than 2 months:* END SURVEY. Please speak to your survey point of contact for further instructions.

5. Please enter the total number of weeks you have spent on TDY/TAD (NOT including training or education) during this assignment:
_____ → *If answer to question 4 minus answer to this question equals less than 2 months:* END SURVEY. Please speak to your survey point of contact for further instructions.

6. Please enter the total number of weeks you have spent on TDY/TAD due to training or education, during this assignment:
_____ → *If answer to question 4 minus answer to this question equals less than 2 months:* END SURVEY. Please speak to your survey point of contact for further instructions.

7. What is your pay grade?
❏ O-1
❏ O-2
❏ O-3

❏ O-4
❏ O-5
❏ O-6
❏ O-7
❏ O-8
❏ O-9
❏ O-10

8. What is your gender?
❏ Male
❏ Female

9. What is your race?
❏ White
❏ Black or African American
❏ Native American or Alaska Native
❏ Asian (e.g., Asian Indian, Chinese, Filipino, Japanese, Korean, Vietnamese)
❏ Native Hawaiian or Other Pacific Islander
❏ Other

10. How many years have you been a commissioned officer?

_____

11. Are you:
❏ Active Duty List Officer?
❏ Reserve Component Officer?

12. Of the following list, what intermediate or senior level schools have you completed? Next to each school you select, note your pay grade at time of completion.

NOTE TO PROGRAMMER: HAVE PULL-DOWNS TO RIGHT OF SCHOOLS FOR THEM TO IDENTIFY PAY GRADE.

Joint and Combined Schools
❏ National War College (grade choices O-1–O-10)

❏ Industrial College of the Armed Forces (grade choices O-1–O-10)
❏ School of Information Warfare and Strategy (grade choices O-1–O-10)
❏ Joint Forces Staff College (grade choices O-1–O-10)
❏ Joint and Combined Staff Officer School (grade choices O-1–O-10)
❏ Joint and Combined Warfighting School (grade choices O-1–O-10)

U.S. Army Service Schools
❏ U.S. Army War College (resident) (grade choices O-1–O-10)
❏ U.S. Army War College (nonresident) (grade choices O-1–O-10)
❏ Army Command and General Staff College (resident) (grade choices O-1–O-10)
❏ Army Command and General Staff College (nonresident) (grade choices O-1–O-10)

U.S. Navy Service Schools
❏ College of Naval Warfare (grade choices O-1–O-10)
❏ College of Naval Command and Staff (resident) (grade choices O-1–O-10)
❏ College of Continuing Education (Navy Intermediate Level College nonresident) (grade choices O-1–O-10)
❏ Naval Postgraduate School (grade choices O-1–O-10)

U.S. Air Force Service Schools
❏ Air War College (grade choices O-1–O-10)
❏ Air Command and Staff College (resident) (grade choices O-1–O-10)
❏ Air Command and Staff College (nonresident) (grade choices O-1–O-10)

U.S. Marine Corps Service Schools
❏ Marine Corps War College (grade choices O-1–O-10)
❏ Marine Corps Command and Staff College (grade choices O-1–O-10)

❏ Marine Corps College of Continuing Education (grade choices O-1–O-10)

13. If you have received credit for postgraduate education at another civilian or military institution not listed above (to include international institutions), please enter the name of the institution and the pay grade you held when you graduated in the boxes below.

Institution	Grade

14. Have you received credit for JPME I?
❏ Yes
❏ No

15. Have you received credit for JPME II?
❏ Yes
❏ No

16. What is your service and component?
❏ U.S. Army
❏ U.S. Army National Guard
❏ U.S. Army Reserve
❏ U.S. Navy
❏ U.S. Naval Reserve
❏ U.S. Air Force
❏ U.S. Air National Guard
❏ U.S. Air Force Reserve
❏ U.S. Marine Corps
❏ U.S. Marine Corps Reserve

If answer to question 16 is U.S. Air Force, U.S. Air National Guard, or U.S. Air Force Reserve:

17. Enter your 4-digit AFSC, e.g., 11M3 (qualified Mobility Pilot). If unsure of 4-digit AFSC, enter 2-digit career area code.

———————

If answer to question 16 is U.S. Army, U.S. Army National Guard, or U.S. Army Reserve:

18. Enter your numeric Area of Concentration (AOC) code, e.g., 11A (Infantry). If you are unsure of your AOC, enter your 2-digit branch or functional area code.

———————

If answer to question 16 is U.S. Navy, U.S. Naval Reserve:

19. Enter your 4-digit officer designator code, e.g., 1110 (Surface Warfare Officer).

———————

If answer to question 16 is U.S. Marine Corps, U.S. Marine Corps Reserve:

20. Enter your 4-digit Military Occupational Specialty (MOS) code, e.g., 0802 (Field Artillery Officer). If you are unsure of your MOS, enter your 2-digit occupational field code.

———————

21. Are you a Joint Specialty Officer (JSO)?
❑ Yes
❑ No
❑ Unsure

NOTE TO PROGRAMMER: THE NEXT THREE QUESTIONS SHOULD BE IN A SINGLE TABLE, WITH THE SAME OPTIONS FOR EACH QUESTION, AND THE QUESTION VERSIONS VARY BY SERVICE, BASED ON PARTICIPANTS ANSWER TO #16.

22. Who is your 1st-level supervisor (e.g., Rater or Reporting Senior)?

23. Who is your 2nd-level supervisor (e.g., Intermediate Rater, Additional Rater, or Reviewing Senior)?

24. Who is your 3rd-level supervisor (e.g., Senior Rater or Reviewer)?

NOTE TO PROGRAMMER: OPTIONS FOR QUESTIONS 22–24:
- ❏ U.S. Army Officer
- ❏ U.S. Army National Guard Officer
- ❏ U.S. Army Reserve Officer
- ❏ U.S. Navy Officer
- ❏ U.S. Naval Reserve Officer
- ❏ U.S Air Force Officer
- ❏ U.S. Air National Guard Officer
- ❏ U.S. Air Force Reserve Officer
- ❏ U.S. Marine Corps Officer
- ❏ U.S. Marine Corps Reserve Officer
- ❏ Non-U.S. Military Officer
- ❏ DOD Civilian
- ❏ Other U.S. Civilian
- ❏ Non-U.S. Civilian
- ❏ Not applicable

25. In your current position, do you serve full time with another service?
- ❏ Yes → 25a. Are you assigned to a billet in that other service?
  - ❏ Yes ❏ No
- ❏ No → SKIP TO 26

26. Do you serve full time with the armed forces of another nation or with an international military or treaty organization (e.g., a U.S. officer assigned to a billet in the headquarters of NATO, a liaison officer at the headquarters of a foreign military service, an officer assigned full time to an element of the United Nations), and are you formally assigned to a billet in that organization?
- ❏ Yes
- ❏ No

27. Are you assigned to both your own service and a joint, combined, or international organization? (Example, an officer assigned to a billet in the G3, Eighth U.S. Army, while simultaneously assigned to positions in the J3, U.S. Forces Korea, and the C3, Combined Forces Command [ROK/U.S.]). Such assignments are referred to as "Dual Hat Positions."
❏ Yes
❏ No

28. Do you serve in Joint Task Force Headquarters Staff?
❏ Yes
❏ No

29. Do you serve in a Joint Task Force Subordinate Organization?
❏ Yes
❏ No

30. Do you serve in a Joint Task Force Service Component?
❏ Yes → 30a. Are you permanently assigned to it? ❏ Yes ❏ No
❏ No → SKIP TO 32.

31. NOTE: No question 31 in incumbent version.

32. Where is the billet located?
❏ United States (including Alaska and Hawaii)
❏ Iraq
❏ Other Middle East
❏ South Asia (e.g., Pakistan, Afghanistan)
❏ Korea
❏ Cuba
❏ Europe
❏ Other nation outside of the United States
❏ Afloat at sea

33. Are you currently serving at your home base?
❏ Yes
❏ No

34. Are you currently receiving Family Separation Allowance (FSA), or would you collect FSA in this position if you had dependents?
❑ Yes
❑ No

35. Are you currently receiving Special Pay for Duty Subject to Hostile Fire or Imminent Danger?
❑ Yes
❑ No

36. Are you currently receiving Special Pay for Hardship Duty?
❑ Yes
❑ No

37. Is your pay subject to Combat Zone Tax Exclusion?
❑ Yes
❑ No

38. Please indicate the approximate percentage of your work time you spend reviewing or deciding matters yourself, as opposed to preparing others to review or decide matters.

NOTE TO PROGRAMMER: PERMIT PARTICIPANT TO FILL IN BLANK NEXT TO "REVIEWING/DECIDING MATTERS MYSELF." SURVEY SHOULD AUTOMATICALLY CALCULATE THE COMPLEMENTARY PERCENTAGE TO FILL IN THE BLANK NEXT TO "PREPARING OTHERS TO REVIEW/ DECIDE MATTERS." PERMIT PARTICIPANT TO REVISE NUMBERS, IF NECESSARY.
Reviewing/deciding matters myself _____%
Preparing others to review/decide matters _____%

39. Indicate which one of the following statements best describes the primary focus of your efforts (mark one response).

❏ The primary focus of my efforts is on operational or supportability matters pertaining to a Combatant Commander's Area of Responsibility (AOR) or several AORs.

❏ The primary focus of my efforts is on defensewide issues or matters that affect one or more Combatant Commanders, Military Departments, or Defense Agencies.

❏ None of the above.

40. If you could choose only one of the following, which best summarizes the level of your job (please click on the hyperlinked choices if you are unsure)?

❏ Strategic

❏ Operational

❏ Tactical

41. On average, how many hours per week do you work?

————————

42. Select the tasks you typically perform. Please select all that apply.

❏ Provide strategic direction and integration

❏ Legal affairs

❏ Inspector General activities

❏ Conduct mobilization

❏ Provide administrative or technical support

❏ Develop, conduct, or provide intelligence, surveillance, and reconnaissance

❏ Provide or exercise command and control

❏ Employ forces

❏ Employ firepower or other assets

❏ Deploy and maneuver forces

❏ Provide or coordinate protection of the force, or protect the force

❏ Special operations

❏ Conduct deployment, redeployment, movement, or maneuver of forces

❏ Counter or manage deterrence of CBRNE weapons, or operate in a CBRNE environment

- ❏ Mapping, charting, and geodesy
- ❏ Provide sustainment
- ❏ Provide logistics or combat service support
- ❏ Combat engineering
- ❏ Maintenance
- ❏ Industrial management
- ❏ Engineering
- ❏ Civil affairs and psychological operations
- ❏ Coordinate counter-proliferation in theater
- ❏ Foster multinational, interagency, alliance, or regional relations
- ❏ Host nation security
- ❏ Targeting of enemy information systems
- ❏ Sustain theater forces' command, control, communications, and computers (C4)
- ❏ Develop/assess joint doctrine
- ❏ Develop/assess joint policies
- ❏ Establish theater force requirements and readiness
- ❏ Resource/financial management
- ❏ Medical/health services
- ❏ Research, development, testing, evaluation, & simulations
- ❏ Conduct force development
- ❏ Operations other than war
- ❏ Law enforcement
- ❏ Safety

43. The tasks you chose in the previous question will appear below. Enter the number of hours per week you perform each task.

Tasks	Hours

44. The tasks you chose in the previous question will appear below. Next to each task is a percentage reflecting the percentage of time

during the week you spend performing the task. This is based on your answer to the previous question, as well as your answer to question 41 regarding the total number of hours per week you work. Do the percentages accurately reflect how you spend your time during the week?

Tasks	Hours

NOTE TO PROGRAMMER: PARTICIPANTS SHOULD BE GIVEN AN OPPORTUNITY TO RETURN TO #44 AND REVISE THEIR HOURS WORKED FOR EACH TASK OR RETURN TO #41 AND REVISE THEIR TOTAL HOURS WORKED.

45. For each of your identified tasks, please select the relative level of importance to your job.

NOTE TO PROGRAMMER: LIST SELECTED TASKS AND THEN HAVE THESE OPTIONS AS PULL-DOWNS NEXT TO EACH TASK. ONLY PERMIT ONE TASK TO BE SELECTED AS THE MOST IMPORTANT THING.

The most important thing I do (note that you can only select one task as most important)
Primary to what I do
Secondary to what I do
Peripheral/less important to what I do

46. Please select the level of responsibility you hold for each of the tasks you perform.

NOTE TO PROGRAMMER: LIST TASKS AND THEN HAVE THESE OPTIONS AS PULL-DOWNS NEXT TO EACH TASK.

minimal responsibility
equally shared responsibility
mostly responsible
sole responsibility

47. With what organizations do you interact? Select all that apply. For each organization selected, please quantify the frequency of your interaction.

NOTE TO PROGRAMMER: NEXT TO EACH ORGANIZATION, HAVE PULL-DOWN MENU WITH FOLLOWING CHOICES.

Daily
Multiple times a week
Weekly
Multiple times a month
Monthly
Multiple times a year
Once a year
Less than once a year

❑ DOD—Office of the Secretary of Defense	❑ DOD—Joint Chiefs of Staff	❑ DOD—U.S. Army
❑ DOD—U.S. Army National Guard	❑ DOD—U.S. Army Reserve	❑ DOD—U.S. Navy
❑ DOD—U.S. Naval Reserve	❑ DOD—U.S. Air Force	❑ DOD—U.S. Air National Guard
❑ DOD—U.S. Air Force Reserve	❑ DOD—U.S. Marine Corps	❑ DOD—U.S. Marine Corps Reserve
❑ DOD—CENTCOM	❑ DOD—EUCOM	❑ DOD—JFCOM
❑ DOD—NORTHCOM	❑ DOD—PACOM	❑ DOD—SOCOM
❑ DOD—SOUTHCOM	❑ DOD—SPACECOM	❑ DOD—STRATCOM
❑ DOD—TRANSCOM	❑ DOD—Industrial College of the Armed Forces	❑ DOD—Information Resource Management College
❑ DOD—Joint Forces Staff College	❑ DOD—Joint Military Intelligence College	❑ DOD—National Defense University

❑ DOD—Army Research Laboratory	❑ DOD—Defense Advanced Research Projects Agency	❑ DOD—National Reconnaissance Office
❑ DOD—Defense Intelligence Agency	❑ DOD—Defense Logistics Agency	❑ DOD—Department of Defense Field Activities
❑ DOD—Defense Threat Reduction Agency	❑ DOD—Missile Defense Agency	❑ DOD—Defense Security Cooperation Agency
❑ DOD—National Geospatial-Intelligence Agency (formerly NIMA)	❑ DOD—National Security Agency—Central Security Service	❑ DOD—Defense Commissary Agency
❑ DOD—Defense Contract Audit Agency	❑ DOD—Defense Contract Management Agency	❑ DOD—Defense Finance and Accounting Service
❑ DOD—Defense Legal Services Agency	❑ DOD—Defense Information Systems Agency	❑ DOD—DOD Computer Emergency Response Team
❑ DHS—Bureau of Customs and Border Protection	❑ DHS—Bureau of Immigration and Customs Enforcement	❑ DHS—Emergency Preparedness & Response Directorate
❑ DHS—Federal Emergency Management Agency	❑ DHS—Federal Law Enforcement Training Center	❑ DHS—Transportation Security Administration
❑ DHS—U.S. Coast Guard	❑ DHS—U.S. Secret Service	❑ DHS—Other
❑ Central Intelligence Agency	❑ Other independent agency or government corporation	❑ Executive Branch
❑ Legislative Branch	❑ Judicial Branch	❑ Department of Agriculture
❑ Department of Commerce	❑ Department of Interior	❑ Department of Justice
❑ Department of State	❑ Department of Transportation	❑ Department of the Treasury
❑ Department of Energy	❑ Department of Health and Human Services	❑ The United Nations
❑ Treaty organizations (such as NATO)	❑ U.S. nongovernmental organizations (such as The Red Cross)	❑ Foreign nongovernmental organizations (such as The Red Crescent)
❑ Non-U.S. military		

48. We would like to know who you interact with. Please indicate this information by placing a check in each appropriate box. Check as many as apply and indicate the frequency of interaction.

NOTE TO PROGRAMMER: NEXT TO EACH TYPE OF PERSON, HAVE PULL-DOWN MENU WITH FOLLOWING CHOICES: Daily, Multiple times a week, Weekly, Multiple times a month, Monthly, Multiple times a year, Once a year, Less than once a year.

❏ U.S. Army personnel (Officer, Enlisted, or Civilian; Active Duty, National Guard, or Reserve)

❏ U.S. Navy personnel (Officer, Enlisted or Civilian; Active Duty or Reserve)

❏ U.S. Air Force personnel (Officer, Enlisted or Civilian; Active Duty, National Guard, or Reserve)

❏ U.S. Marine Corps personnel (Officer, Enlisted or Civilian; Active Duty or Reserve)

❏ Other DOD Civilian

❏ Other U.S. Civilian

❏ Non-U.S. Civilian

❏ Non-U.S. Military Officer

❏ Not applicable

49. In this section, we would like to know two things:
   a. What knowledge you feel was required or helpful for this position, and
   b. In what knowledge you have gained—or expect to gain—either familiarity or proficiency while in this position.

   There are two sets of columns by each type of knowledge listed below. The first set of columns permits you to identify the type of knowledge that is either "required" or "helpful" for someone serving in your position. Please select buttons from the second set of columns to identify those areas of expertise in which you will gain either proficiency or familiarity while serving in this assignment. For each subject, you will be able to select only one button from each set of columns.

Answer List "C"

	For this position, I find this knowledge ___ :		While in this position, I have (or expect to) become ___ with this knowledge:	
	Required	Helpful	Proficient	Familiar
National Military Capabilities, Organization, and Command Structure				
Roles, relationships, and functions of the NCA, JCS, CoComs, NSC, JFC, and the CJCS	□	□	□	□
Force structure requirements and resultant capabilities and limitations of U.S. military forces	□	□	□	□
How the U.S. military plans, executes, and trains for joint, interagency, and multinational ops	□	□	□	□
Service-unique capability, limitation, doctrine, and command structure integration	□	□	□	□
National Military Strategy				
Art and science of developing, deploying, employing, and sustaining the military resources	□	□	□	□
Capabilities and limitations of the U.S. force structure and their effect on joint military strategy	□	□	□	□
Concepts of the strategic decisionmaking and defense planning processes	□	□	□	□
Resource needs, both national and international, for national defense	□	□	□	□
Key considerations that shape the development of national military strategy	□	□	□	□
Current National Military Strategy and other examples of U.S. and foreign military strategies	□	□	□	□
DoD long-term and immediate process for strategic planning and assessment	□	□	□	□
National Security Strategy				
National security policy process, to include the integration of the instruments of national power	□	□	□	□
Impact of defense acquisition and its implications for enhancing our joint military capabilities	□	□	□	□
Relationships between the military, Congress, NSC, DoD agencies, and the public	□	□	□	□
Developing, applying, and coordinating the instruments of national power	□	□	□	□
How national policy is turned into executable military strategies	□	□	□	□
Capabilities and vulnerabilities of U.S. industry and infrastructure in a global market	□	□	□	□
National security technological environment for current and future competitive advantage	□	□	□	□

Answer List "C"—continued

	Required	Helpful	Proficient	Familiar
National Security Policy Process				
Origins, responsibilities, organization, and modus operandi of the NSC system	❏	❏	❏	❏
How major governmental and NGOs influence and implement national security policies	❏	❏	❏	❏
How the U.S. government prioritizes among issues for developing national-level strategies	❏	❏	❏	❏
National Planning Systems and Processes				
National security decisionmaking system and the policy formulation process	❏	❏	❏	❏
Responsibilities and relationships of the interagency and the joint community	❏	❏	❏	❏
DoD processes by which national ends, ways, and means are reconciled, integrated, and applied	❏	❏	❏	❏
How time, coordination, policy, politics, doctrine, and national power affect the planning process	❏	❏	❏	❏
Command, Control, Communications, Computers, Intelligence, Surveillance, and Reconnaissance (C4ISR)				
How C4ISR systems apply at the tactical and operational levels of war	❏	❏	❏	❏
How IO is incorporated into both the deliberate and crisis action planning processes	❏	❏	❏	❏
How opportunities and vulnerabilities are created by increased reliance on IT	❏	❏	❏	❏
Integrating IO and C4 to support the National Military and National Security Strategies	❏	❏	❏	❏
Integrating IO and C4 into the theater and strategic campaign development process	❏	❏	❏	❏
IO, IW, and C4I concepts in joint operations	❏	❏	❏	❏
Theater Strategy and Campaigning				
Role of the unified commander in developing theater plans, policies, and strategies	❏	❏	❏	❏
Coordination of U.S. military plans/actions with foreign forces, interagency and NGOs	❏	❏	❏	❏
How joint and multinational campaigns and operations support national objectives	❏	❏	❏	❏
Combatant Commander's perspective of the resources required to support campaign plans	❏	❏	❏	❏
Organization, responsibilities, and capabilities of military forces available to the JFCs	❏	❏	❏	❏

Answer List "C"—continued

Required	Helpful	Proficient	Familiar	
				Geo-Strategic Context
☐	☐	☐	☐	Current social, cultural, political, economic, military, technological, and historical issues
☐	☐	☐	☐	Roles and influence of international organizations and other non-state actors
☐	☐	☐	☐	Key military, nonmilitary, and transnational challenges to U.S. national security
				Instruments of National Power
☐	☐	☐	☐	Fundamental characteristics, capabilities, and limitations instruments of national power
☐	☐	☐	☐	Employment of diplomatic, economic, military, and informational instruments of national power
				Joint Operational Art
☐	☐	☐	☐	Joint doctrine and the joint operational art
☐	☐	☐	☐	Integration of service, joint, interagency, and multinational capabilities
				Joint Warfare Fundamentals
☐	☐	☐	☐	Each combatant command's mission, organization, and responsibilities
☐	☐	☐	☐	Joint aspects of military operations other than war (MOOTW)
☐	☐	☐	☐	Capabilities of other services' weapon systems
				Joint Campaigning
☐	☐	☐	☐	JTF organization, including who can form a JTF and how and when a JTF is formed
☐	☐	☐	☐	Characteristics of a joint campaign and the relationships of supporting capabilities
				Joint Doctrine
☐	☐	☐	☐	Current joint doctrine
☐	☐	☐	☐	Factors influencing joint doctrine
☐	☐	☐	☐	Relationship between service doctrine and joint doctrine
				Joint and Multinational Forces at the Operational Level of War
☐	☐	☐	☐	Considerations for employing joint and multinational forces at the operational level of war

Answer List "C"—continued

Required	Helpful	Proficient	Familiar	
				Joint and Multinational Forces at the Operational Level of War (cont.)
☐	☐	☐	☐	How theory and principles of war apply at the operational level of war
☐	☐	☐	☐	Relationships among national objectives, military objectives, and conflict termination
☐	☐	☐	☐	Relationships among the strategic, operational, and tactical levels of war
				Joint Planning and Execution Processes
☐	☐	☐	☐	Relationship between national objectives and means availability
☐	☐	☐	☐	Effect of time, coordination, policy changes, and political developments on the planning process
☐	☐	☐	☐	How national, joint, and service intelligence organizations support JFCs
☐	☐	☐	☐	Integrating battle space support systems into campaign/theater planning and operations
				Others
☐	☐	☐	☐	Inspector General activities, legal/legislative, law enforcement, physical security or investigations
☐	☐	☐	☐	Special operations, operations other than war, tactical matters (i.e., training exercises, etc.)
☐	☐	☐	☐	Manpower/personnel, training, education, logistics, acquisition, or general administration
☐	☐	☐	☐	R&D, engineering, scientific matters (includes weather, environment, etc.), CBRNE matters
☐	☐	☐	☐	Medical or health services

50. As a result of current events, my experience in this position was different from that of my predecessors.
❏ Strongly agree
❏ Agree
❏ Neither agree nor disagree
❏ Disagree
❏ Strongly disagree
❏ Not applicable

51. My assessment of this position depends upon current events, making it unlikely that future occupants will have the same experience.
❏ Strongly agree
❏ Agree
❏ Neither agree nor disagree
❏ Disagree
❏ Strongly disagree
❏ Not applicable

52. This position gives me significant experience in multiservice matters.
❏ Strongly agree
❏ Agree
❏ Neither agree nor disagree
❏ Disagree
❏ Strongly disagree
❏ Not applicable

53. This position gives me significant experience in multinational matters.
❏ Strongly agree
❏ Agree
❏ Neither agree nor disagree
❏ Disagree
❏ Strongly disagree
❏ Not applicable

54. This position gives me significant experience in interagency matters.
- ❑ Strongly agree
- ❑ Agree
- ❑ Neither agree nor disagree
- ❑ Disagree
- ❑ Strongly disagree
- ❑ Not applicable

55. In order to perform my duties successfully, I have found JPME I
- ❑ Required
- ❑ Desired
- ❑ Not helpful

56. In order to perform my duties successfully, I have found JPME II
- ❑ Required
- ❑ Desired
- ❑ Not helpful

57. In order to perform my duties successfully, I have found joint training or education (other than JPME)
- ❑ Required
- ❑ Desired
- ❑ Not helpful

58. In order to perform my duties successfully, I have found prior experience in a joint environment
- ❑ Required
- ❑ Desired
- ❑ Not helpful

59. To what extent do you draw upon your primary military specialty (i.e., AOC code, MOS, AFSC, or Navy designator) to perform in this position?
- ❑ All of the time
- ❑ Most of the time

❑ Half of the time
❑ Some of the time
❑ Not at all
❑ Not applicable

60. To what extent do you draw upon knowledge of your service's capabilities to perform in this position?
❑ All of the time
❑ Most of the time
❑ Half of the time
❑ Some of the time
❑ Not at all
❑ Not applicable

61. How many weeks did it take in this position to become comfortable in a joint environment?
❑ _____
❑ Not yet comfortable
❑ Not in a joint environment
❑ Not applicable for other reasons

62. What is the planned length of your current assignment (in months)? _____

63. How many months do you think your assignment should last?
_____

64. How many months do you think a typical joint duty assignment should last? _____

65. Which of the following was most important to you in this assignment?
❑ Service core competencies
❑ Prior joint experience
❑ Specialized training and orientation in joint matters
❑ Functional expertise, e.g., operations, intelligence

❑ Other not listed here—please specify:
❑ Not applicable

66. In your opinion, what is the most important thing your successor should possess?
❑ Service core competencies
❑ Prior joint experience
❑ Specialized training and orientation in joint matters
❑ Functional expertise, e.g., operations, intelligence
❑ Other not listed here—please specify:
❑ Not applicable

67. A civilian could perform the duties and responsibilities of this position just as effectively.
❑ Strongly agree
❑ Agree
❑ Neither agree nor disagree
❑ Disagree
❑ Strongly disagree
❑ Not applicable

68. My position requires unique knowledge of my own service and could not be performed by an officer of another service
❑ Strongly agree
❑ Agree
❑ Neither agree nor disagree
❑ Disagree
❑ Strongly disagree
❑ Not applicable

69. Morale problems will exist if joint duty credit is awarded for some positions in my immediate organization but not for others.
❑ Strongly agree
❑ Agree
❑ Neither agree nor disagree
❑ Disagree

❏ Strongly disagree
❏ Not applicable

NOTE TO PROGRAMMER: IF ANSWER TO #5 WAS EQUAL TO OR GREATER THAN 3 MONTHS FOR EACH YEAR SPENT IN THE ASSIGNMENT, THEN ASK THIS QUESTION:

70. How much of your assessment of this billet is based on experience gained through TAD/TDY?

❏ Considerable amount
❏ Moderate amount
❏ Minimal amount
❏ Not at all
❏ Not applicable

71. Is there anything else you would like to tell us? _____

Protocol Version for Individual Other Than Position Incumbent

1. Please enter the position number and password below:
   Position number _____
   Password _____

2. Is this the billet you have been asked to evaluate?
   ❑ Yes →
   ❑ No → END SURVEY. Please speak to your survey point of contact for further instructions.

3. Are you the
   ❑ Person occupying the billet?
   ❑ Supervisor of the billet?
   ❑ Another person designated to complete the survey?

NOTE TO PROGRAMMER: IF NOT PERSON OCCUPYING BILLET, CONTINUE WITH THIS PROTOCOL VERSION.

4. Please enter your nonincumbent code.
   If you have evaluated (or will be evaluating) your own positions, enter your own position code as the nonincumbent code below.
   If you will not also be evaluating your own position, you should have been given a nonincumbent code by your survey POC. If you did not receive one, please contact your survey POC to obtain one.
   Nonincumbent code _____

5. How many months have you been familiar with this billet?
   _____ → *If less than 2 months:* END SURVEY. Please speak to your survey point of contact for further instructions.

6. NOTE: No 6 in nonincumbent version.

The next series of questions refer to you, the person completing this survey as a nonincumbant.

7. What is your pay grade?
- ❑ O-1
- ❑ O-2
- ❑ O-3
- ❑ O-4
- ❑ O-5
- ❑ O-6
- ❑ O-7
- ❑ O-8
- ❑ O-9
- ❑ O-10

8. What is your gender?
- ❑ Male
- ❑ Female

9. What is your race?
- ❑ White
- ❑ Black or African American
- ❑ Indian or Alaska Native
- ❑ Asian (e.g., Asian Indian, Chinese, Filipino, Japanese, Korean, Vietnamese)
- ❑ Native Hawaiian or Other Pacific Islander (e.g., Samoan, Guamanian or Chamorro)
- ❑ Other

10. How many years have you been a commissioned officer?

_____

11. Are you:
- ❑ Active Duty List Officer?
- ❑ Reserve Component Officer?

12. Of the following list, what intermediate or senior level schools have you completed? Next to each school you select, note your pay grade at time of completion.

Joint and Combined Schools
- ❏ National War College (grade choices O-1–O-10)
- ❏ Industrial College of the Armed Forces (grade choices O-1–O-10)
- ❏ School of Information Warfare and Strategy (grade choices O-1–O-10)
- ❏ Joint Forces Staff College (grade choices O-1–O-10)
- ❏ Joint and Combined Staff Officer School (grade choices O-1–O-10)
- ❏ Joint and Combined Warfighting School (grade choices O-1–O-10)

U.S. Army Service Schools
- ❏ U.S. Army War College (resident) (grade choices O-1–O-10)
- ❏ U.S. Army War College (nonresident) (grade choices O-1–O-10)
- ❏ Army Command and General Staff College (resident) (grade choices O-1–O-10)
- ❏ Army Command and General Staff College (nonresident) (grade choices O-1–O-10)

U.S. Navy Service Schools
- ❏ College of Naval Warfare (grade choices O-1–O-10)
- ❏ College of Naval Command and Staff (resident) (grade choices O-1–O-10)
- ❏ College of Continuing Education (Navy Intermediate Level College nonresident) (grade choices O-1–O-10)
- ❏ Naval Postgraduate School (grade choices O-1–O-10)

U.S. Air Force Service Schools
- ❏ Air War College (grade choices O-1–O-10)
- ❏ Air Command and Staff College (resident) (grade choices O-1–O-10)
- ❏ Air Command and Staff College (nonresident) (grade choices O-1–O-10)

U.S. Marine Corps Service Schools
- ❏ Marine Corps War College (grade choices O-1–O-10)

❏ Marine Corps Command and Staff College (grade choices O-1–O-10)
❏ Marine Corps College of Continuing Education (grade choices O-1–O-10)

13. If you have received credit for postgraduate education at another civilian or military institution not listed above (to include international institutions), please enter the name of the institution and the pay grade you held when you graduated in the boxes below.

Institution	Grade

14. Have you received credit for JPME I?
❏ Yes
❏ No

15. Have you received credit for JPME II?
❏ Yes
❏ No

16. What is your service and component?
❏ U.S. Army
❏ U.S. Army National Guard
❏ U.S. Army Reserve
❏ U.S. Navy
❏ U.S. Naval Reserve
❏ U.S. Air Force
❏ U.S. Air National Guard
❏ U.S. Air Force Reserve
❏ U.S. Marine Corps
❏ U.S. Marine Corps Reserve

NOTE TO PROGRAMMER: NO QUESTIONS 17–20 IN NON-INCUMBENT VERSION.

21. Are you a Joint Specialty Officer (JSO)?
- ❏ Yes
- ❏ No
- ❏ Unsure

The remainder of the questions in this survey refer to the billet you are evaluating.

NOTE TO PROGRAMMER: NEXT THREE QUESTIONS SHOULD BE IN A SINGLE TABLE, WITH THE SAME OPTIONS FOR EACH QUESTION.

22. Who is the 1st-level supervisor (e.g., Rater or Reporting Senior) for this position?

23. Who is the 2nd-level supervisor (e.g., Intermediate Rater, Additional Rater or Reviewing Senior) for this position?

24. Who is the 3rd-level supervisor (e.g., Senior Rater or Reviewer) for this position?

NOTE TO PROGRAMMER: OPTIONS FOR QUESTIONS 22–24:
- ❏ U.S. Army Officer
- ❏ U.S. Army National Guard Officer
- ❏ U.S. Army Reserve Officer
- ❏ U.S. Navy Officer
- ❏ U.S. Naval Reserve Officer
- ❏ U.S. Air Force Officer
- ❏ U.S. Air National Guard Officer
- ❏ U.S. Air Force Reserve Officer
- ❏ U.S. Marine Corps Officer
- ❏ U.S. Marine Corps Reserve Officer
- ❏ Non-U.S. Military Officer
- ❏ DOD Civilian

❑ Other U.S. Civilian
❑ Non-U.S. Civilian
❑ Not applicable

25. Would a person in this assignment typically serve full time with another service?
❑ Yes → 25b. Would that person be assigned to a billet in that other service? ❑ Yes ❑ No
❑ No → SKIP TO 26

26. Would a person in this assignment typically serve full time with the armed forces of another nation or with an international military or treaty organization (e.g., a U.S. officer assigned to a assignment in the headquarters of NATO, a liaison officer at the headquarters of a foreign military service, an officer assigned full time to an element of the United Nations) and be formally assigned to a billet in that organization?
❑ Yes
❑ No

27. Would a person in this assignment typically be assigned to both his/her own service and a joint, combined, or international organization? (Example, an officer assigned to a assignment in the G3, Eighth U.S. Army, while simultaneously assigned to positions in the J3, U.S. Forces Korea, and the C3, Combined Forces Command [ROK/U.S.].) Such assignments are referred to as "Dual Hat Positions."
❑ Yes
❑ No

28. Would a person in this assignment typically serve in Joint Task Force Headquarters Staff?
❑ Yes
❑ No

29. Would a person in this assignment typically serve in a Joint Task Force Subordinate Organization?
- ❏ Yes
- ❏ No

30. Would a person in this assignment typically serve in a Joint Task Force Service Component?
- ❏ Yes → 30a. Would that person be permanently assigned to it?
  ❏ Yes ❏ No
- ❏ No → SKIP TO 31.

31. Would a person in this assignment typically be a reservist temporarily on active duty?
- ❏ Yes
- ❏ No

32. Where is the billet located?
- ❏ United States (including Alaska and Hawaii)
- ❏ Iraq
- ❏ Other Middle East
- ❏ South Asia (e.g., Pakistan, Afghanistan)
- ❏ Korea
- ❏ Cuba
- ❏ Europe
- ❏ Other nation outside of the United States
- ❏ Afloat at sea

33. Would a person in this position typically be serving at his/her home base?
- ❏ Yes
- ❏ No

34. Would a person in this position typically receive Family Separation Allowance (FSA), if they had dependents?
- ❏ Yes
- ❏ No

35. Would a person in this position typically receive Special Pay for Duty Subject to Hostile Fire or Imminent Danger?
❑ Yes
❑ No

36. Would a person in this position typically receive Special Pay for Hardship Duty?
❑ Yes
❑ No

37. Is your pay subject to Combat Zone Tax Exclusion?
❑ Yes
❑ No

38. Please indicate the approximate percentage of time worked a person in this position would review or decide matters themselves, or prepare others to review or decide matters.

NOTE TO PROGRAMMER: PERMIT PARTICIPANT TO FILL IN BLANK NEXT TO "REVIEWING/DECIDING MATTERS MYSELF." SURVEY SHOULD AUTOMATICALLY CALCULATE THE COMPLEMENTARY PERCENTAGE TO FILL IN THE BLANK NEXT TO "PREPARING OTHERS TO REVIEW/DECIDE MATTERS." PERMIT PARTICIPANT TO REVISE NUMBERS, IF NECESSARY.

Review/decide matters themselves _____%
Prepare others to review/decide matters _____%

39. Indicate which one of the following statements best describes the primary focus of the efforts of a person in this position (mark one response).
❑ The primary focus of the efforts of a person in this position is on operational or supportability matters pertaining to a Combatant Commander's Area of Responsibility (AOR) or several AORs.

❑ The primary focus of the efforts of a person in this position is on defensewide issues or matters that affect one or more Combatant Commanders, Military Departments, or Defense Agencies.

❑ None of the above.

40. If you could choose only one of the following, which best summarizes the level of this job?

❑ Strategic

❑ Operational

❑ Tactical

41. On average, how many hours per week would a person in this position work? _____

42. Select the tasks a person in this position would typically perform. Please select all that apply.

❑ Provide strategic direction and integration

❑ Legal affairs

❑ Inspector General activities

❑ Conduct mobilization

❑ Provide administrative or technical support

❑ Develop, conduct, or provide intelligence, surveillance, and reconnaissance

❑ Provide or exercise command and control

❑ Employ forces

❑ Employ firepower or other assets

❑ Deploy and maneuver forces

❑ Provide or coordinate protection of the force, or protect the force

❑ Special operations

❑ Conduct deployment, redeployment, movement, or maneuver of forces

❑ Counter or manage deterrence of CBRNE weapons, or operate in a CBRNE environment

❑ Mapping, charting and geodesy

❑ Provide sustainment

❑ Provide logistics or combat service support

❑ Combat engineering
❑ Maintenance
❑ Industrial management
❑ Engineering
❑ Civil affairs and psychological operations
❑ Coordinate counter-proliferation in theater
❑ Foster multinational, interagency, alliance, or regional relations
❑ Host nation security
❑ Targeting of enemy information systems
❑ Sustain theater forces' command, control, communications, and computers (C4)
❑ Develop/assess joint doctrine
❑ Develop/assess joint policies
❑ Establish theater force requirements and readiness
❑ Resource/financial management
❑ Medical/health services
❑ Research, development, testing, evaluation, & simulations
❑ Conduct force development
❑ Operations other than war
❑ Law enforcement
❑ Safety

43. The tasks you chose in the previous question will appear below. Enter the number of hours per week a person in this position would typically perform each task.

Tasks	Hours

44. The tasks you chose in the previous question will appear below. Next to each task is a percentage reflecting the percentage of time during the week a person in this position typically spends performing the task. This is based on your answer to the previous

question, as well as your answer to question 41 regarding the total number of hours per week a person in this position works. Do the percentages accurately reflect how a person in this position would typically spend his/her time during the week?

Tasks	Hours

NOTE TO PROGRAMMER: PARTICIPANTS SHOULD BE GIVEN AN OPPORTUNITY TO RETURN TO #44 AND REVISE THEIR HOURS WORKED FOR EACH TASK OR RETURN TO #41 AND REVISE THEIR TOTAL HOURS WORKED.

45. For each of the identified tasks, select the level of importance to this position.

NOTE TO PROGRAMMER, LIST SELECTED TASKS AND THEN HAVE THESE OPTIONS AS PULL-DOWNS NEXT TO EACH TASK. ONLY PERMIT ONE TASK TO BE SELECTED AS THE MOST IMPORTANT THING.

the most important thing a person in this position does (note that you can only select one task as most important)
primary to what they do
secondary to what they do
peripheral/less important to what they do

46. Select the level of responsibility a person in this position would hold for each of the tasks he/she would perform.

NOTE TO PROGRAMMER, LIST TASKS AND THEN HAVE THESE OPTIONS AS PULL-DOWNS NEXT TO EACH TASK.

minimal responsibility
equally shared responsibility
mostly responsible
sole responsibility

47. With what organizations would a person in this position interact? For each organization selected, please quantify the frequency of likely interaction.

NOTE TO PROGRAMMER: NEXT TO EACH ORGANIZATION, HAVE PULL-DOWN MENU WITH FOLLOWING CHOICES:

Daily
Multiple times a week
Weekly
Multiple times a month
Monthly
Multiple times a year
Once a year
Less than once a year

❏ DOD—Office of the Secretary of Defense	❏ DOD—Joint Chiefs of Staff	❏ DOD—U.S. Army
❏ DOD—U.S. Army National Guard	❏ DOD—U.S. Army Reserve	❏ DOD—U.S. Navy
❏ DOD—U.S. Naval Reserve	❏ DOD—U.S. Air Force	❏ DOD—U.S. Air National Guard
❏ DOD—U.S. Air Force Reserve	❏ DOD—U.S. Marine Corps	❏ DOD—U.S. Marine Corps Reserve
❏ DOD—CENTCOM	❏ DOD—EUCOM	❏ DOD—JFCOM
❏ DOD—NORTHCOM	❏ DOD—PACOM	❏ DOD—SOCOM
❏ DOD—SOUTHCOM	❏ DOD—SPACECOM	❏ DOD—STRATCOM
❏ DOD—TRANSCOM	❏ DOD—Industrial College of the Armed Forces	❏ DOD—Information Resource Management College
❏ DOD—Joint Forces Staff College	❏ DOD—Joint Military Intelligence College	❏ DOD—National Defense University
❏ DOD—Army Research Laboratory	❏ DOD—Defense Advanced Research Projects Agency	❏ DOD—National Reconnaissance Office

❑ DOD—Defense Intelligence Agency	❑ DOD—Defense Logistics Agency	❑ DOD—Department of Defense Field Activities
❑ DOD—Defense Threat Reduction Agency	❑ DOD—Missile Defense Agency	❑ DOD—Defense Security Cooperation Agency
❑ DOD—National Geospatial-Intelligence Agency (formerly NIMA)	❑ DOD—National Security Agency—Central Security Service	❑ DOD—Defense Commissary Agency
❑ DOD—Defense Contract Audit Agency	❑ DOD—Defense Contract Management Agency	❑ DOD—Defense Finance and Accounting Service
❑ DOD—Defense Legal Services Agency	❑ DOD—Defense Information Systems Agency	❑ DOD—DOD Computer Emergency Response Team
❑ DHS—Bureau of Customs and Border Protection	❑ DHS—Bureau of Immigration and Customs Enforcement	❑ DHS—Emergency Preparedness & Response Directorate
❑ DHS—Federal Emergency Management Agency	❑ DHS—Federal Law Enforcement Training Center	❑ DHS—Transportation Security Administration
❑ DHS—U.S. Coast Guard	❑ DHS—U.S. Secret Service	❑ DHS—Other
❑ Central Intelligence Agency	❑ Other independent agency or government corporation	❑ Executive Branch
❑ Legislative Branch	❑ Judicial Branch	❑ Department of Agriculture
❑ Department of Commerce	❑ Department of Interior	❑ Department of Justice
❑ Department of State	❑ Department of Transportation	❑ Department of the Treasury
❑ Department of Energy	❑ Department of Health and Human Services	❑ The United Nations
❑ Treaty organizations (such as NATO)	❑ U.S. nongovernmental organizations (such as The Red Cross)	❑ Foreign nongovernmental organizations (such as The Red Crescent)
❑ Non-U.S. military		

48. We would like to know who an individual in this position would typically interact with. Please indicate this information by placing a check in each appropriate box. Check as many as apply and select the frequency of likely interaction for those selected.

NOTE TO PROGRAMMER: NEXT TO EACH TYPE OF PERSON, HAVE PULL-DOWN MENU WITH FOLLOWING CHOICES:

Daily

Multiple times a week
Weekly
Multiple times a month
Monthly
Multiple times a year
Once a year
Less than once a year

❏ U.S. Army personnel (Officer, Enlisted, or Civilian; Active Duty, National Guard, or Reserve)
❏ U.S. Navy personnel (Officer, Enlisted, or Civilian; Active Duty or Reserve)
❏ U.S. Air Force personnel (Officer, Enlisted, or Civilian; Active Duty, National Guard, or Reserve)
❏ U.S. Marine Corps personnel (Officer, Enlisted, or Civilian; Active Duty or Reserve)
❏ Other DOD Civilian
❏ Other U.S. Civilian
❏ Non-U.S. Civilian
❏ Non-U.S. Military Officer
❏ Not applicable

49. In this section, we would like to know two things:
  a. What knowledge you feel is required or helpful to a person in this position, and
  b. In what knowledge a person will gain—or can expect to gain—either familiarity or proficiency while in this position.

  There are two sets of columns by each type of knowledge listed below. The first set of columns permits you to identify the type of knowledge that is either "required" or "helpful" for someone serving in this position. Please select buttons from the second set of columns to identify those areas of expertise in which a person would gain either proficiency or familiarity while serving in this assignment. For each subject, you will be able to select only one button from each set of columns.

Answer List "C"

Knowledge	For this position, I find this knowledge: Required	Helpful	While in this position, I have (or expect to) become with this knowledge: Proficient	Familiar
National Military Capabilities, Organization, and Command Structure				
Roles, relationships, and functions of the NCA, JCS, CoComs, NSC, JFC, and the CJCS	☐	☐	☐	☐
Force structure requirements and resultant capabilities and limitations of U.S. military forces	☐	☐	☐	☐
How the U.S. military plans, executes, and trains for joint, interagency, and multinational ops	☐	☐	☐	☐
Service-unique capability, limitation, doctrine, and command structure integration	☐	☐	☐	☐
National Military Strategy				
Art and science of developing, deploying, employing, and sustaining the military resources	☐	☐	☐	☐
Capabilities and limitations of the U.S. force structure and their effect on joint military strategy	☐	☐	☐	☐
Concepts of the strategic decisionmaking and defense planning processes	☐	☐	☐	☐
Resource needs, both national and international, for national defense	☐	☐	☐	☐
Key considerations that shape the development of national military strategy	☐	☐	☐	☐
Current National Military Strategy and other examples of U.S. and foreign military strategies	☐	☐	☐	☐
DoD long-term and immediate process for strategic planning and assessment	☐	☐	☐	☐
National Security Strategy				
National security policy process, to include the integration of the instruments of national power	☐	☐	☐	☐
Impact of defense acquisition and its implications for enhancing our joint military capabilities	☐	☐	☐	☐
Relationships between the military, Congress, NSC, DoD agencies, and the public	☐	☐	☐	☐
Developing, applying, and coordinating the instruments of national power	☐	☐	☐	☐
How national policy is turned into executable military strategies	☐	☐	☐	☐
Capabilities and vulnerabilities of U.S. industry and infrastructure in a global market	☐	☐	☐	☐
National security technological environment for current and future competitive advantage	☐	☐	☐	☐

Answer List "C"—continued

Required	Helpful	Proficient	Familiar	
				National Security Policy Process
❏	❏	❏	❏	Origins, responsibilities, organization, and modus operandi of the NSC system
❏	❏	❏	❏	How major governmental and NGOs influence and implement national security policies
❏	❏	❏	❏	How the U.S. government prioritizes among issues for developing national-level strategies
				National Planning Systems and Processes
❏	❏	❏	❏	National security decisionmaking system and the policy formulation process
❏	❏	❏	❏	Responsibilities and relationships of the interagency and the joint community
❏	❏	❏	❏	DoD processes by which national ends, ways, and means are reconciled, integrated, and applied
❏	❏	❏	❏	How time, coordination, policy, politics, doctrine, and national power affect the planning process
				Command, Control, Communications, Computers, Intelligence, Surveillance, and Reconnaissance (C4ISR)
❏	❏	❏	❏	How C4ISR systems apply at the tactical and operational levels of war
❏	❏	❏	❏	How IO is incorporated into both the deliberate and crisis action planning processes
❏	❏	❏	❏	How opportunities and vulnerabilities are created by increased reliance on IT
❏	❏	❏	❏	Integrating IO and C4 to support the National Military and National Security Strategies
❏	❏	❏	❏	Integrating IO and C4 into the theater and strategic campaign development process
❏	❏	❏	❏	IO, IW, and C4I concepts in joint operations
				Theater Strategy and Campaigning
❏	❏	❏	❏	Role of the unified commander in developing theater plans, policies, and strategies
❏	❏	❏	❏	Coordination of U.S. military plans/actions with foreign forces, interagency and NGOs
❏	❏	❏	❏	How joint and multinational campaigns and operations support national objectives
❏	❏	❏	❏	Combatant Commander's perspective of the resources required to support campaign plans
❏	❏	❏	❏	Organization, responsibilities, and capabilities of military forces available to the JFCs

Answer List "C"—continued

Required	Helpful	Proficient	Familiar	
				Geo-Strategic Context
☐	☐	☐	☐	Current social, cultural, political, economic, military, technological, and historical issues
☐	☐	☐	☐	Roles and influence of international organizations and other non-state actors
☐	☐	☐	☐	Key military, nonmilitary, and transnational challenges to U.S. national security
				Instruments of National Power
☐	☐	☐	☐	Fundamental characteristics, capabilities, and limitations instruments of national power
☐	☐	☐	☐	Employment of diplomatic, economic, military, and informational instruments of national power
				Joint Operational Art
☐	☐	☐	☐	Joint doctrine and the joint operational art
☐	☐	☐	☐	Integration of service, joint, interagency, and multinational capabilities
				Joint Warfare Fundamentals
☐	☐	☐	☐	Each combatant command's mission, organization, and responsibilities
☐	☐	☐	☐	Joint aspects of military operations other than war (MOOTW)
☐	☐	☐	☐	Capabilities of other services' weapon systems
				Joint Campaigning
☐	☐	☐	☐	JTF organization, including who can form a JTF and how and when a JTF is formed
☐	☐	☐	☐	Characteristics of a joint campaign and the relationships of supporting capabilities
				Joint Doctrine
☐	☐	☐	☐	Current joint doctrine
☐	☐	☐	☐	Factors influencing joint doctrine
☐	☐	☐	☐	Relationship between service doctrine and joint doctrine
				Joint and Multinational Forces at the Operational Level of War
☐	☐	☐	☐	Considerations for employing joint and multinational forces at the operational level of war

Answer List "C"—continued

Required	Helpful	Proficient	Familiar	
				Joint and Multinational Forces at the Operational Level of War (cont.)
☐	☐	☐	☐	How theory and principles of war apply at the operational level of war
☐	☐	☐	☐	Relationships among national objectives, military objectives, and conflict termination
☐	☐	☐	☐	Relationships among the strategic, operational, and tactical levels of war
				Joint Planning and Execution Processes
☐	☐	☐	☐	Relationship between national objectives and means availability
☐	☐	☐	☐	Effect of time, coordination, policy changes, and political developments on the planning process
☐	☐	☐	☐	How national, joint, and service intelligence organizations support JFCs
☐	☐	☐	☐	Integrating battle space support systems into campaign/theater planning and operations
				Others
☐	☐	☐	☐	Inspector General activities, legal/legislative, law enforcement, physical security or investigations
☐	☐	☐	☐	Special operations, operations other than war, tactical matters (i.e., training exercises, etc.)
☐	☐	☐	☐	Manpower/personnel, training, education, logistics, acquisition, or general administration
☐	☐	☐	☐	R&D, engineering, scientific matters (includes weather, environment, etc.), CBRNE matters
☐	☐	☐	☐	Medical or health services

50. My evaluation of this billet was affected by current events, and individuals who previously filled this billet had a different experience in this job.
❏ Strongly agree
❏ Agree
❏ Neither agree nor disagree
❏ Disagree
❏ Strongly disagree
❏ Not applicable

51. My assessment of this position depends upon current events, making it unlikely that future occupants will have the experience I have evaluated.
❏ Strongly agree
❏ Agree
❏ Neither agree nor disagree
❏ Disagree
❏ Strongly disagree
❏ Not applicable

52. A person in this position gains significant experience in multi-service matters.
❏ Strongly agree
❏ Agree
❏ Neither agree nor disagree
❏ Disagree
❏ Strongly disagree
❏ Not applicable

53. A person in this position gains significant experience in multi-national matters.
❏ Strongly agree
❏ Agree

❏ Neither agree nor disagree
❏ Disagree
❏ Strongly disagree
❏ Not applicable

54. A person in this position gains significant experience in inter-agency matters.
❏ Strongly agree
❏ Agree
❏ Neither agree nor disagree
❏ Disagree
❏ Strongly disagree
❏ Not applicable

55. In order to perform the duties of this position successfully, an individual would find JPME I
❏ Required
❏ Desired
❏ Not helpful

56. In order to perform the duties of this position successfully, an individual would find JPME II
❏ Required
❏ Desired
❏ Not helpful

57. In order to perform the duties of this position successfully, an individual would find joint training or education (other than JPME)
❏ Required
❏ Desired
❏ Not helpful

58. In order to perform the duties of this position successfully, an individual would find prior experience in a joint environment
❏ Required

❏ Desired
❏ Not helpful

59. To what extent would a person in this position typically draw upon their primary military specialty (i.e., AOC code, MOS, AFSC, or Navy designator) to perform in this position?
❏ Not at all
❏ Some of the time
❏ Half of the time
❏ Most of the time
❏ All of the time
❏ Not sure
❏ Not applicable

60. To what extent would a person in this position typically draw upon knowledge of his/her service's capabilities to perform in this position?
❏ Not at all
❏ Some of the time
❏ Half of the time
❏ Most of the time
❏ All of the time
❏ Not applicable

61. How many weeks do you feel it would take a person in this position to become comfortable in a joint environment?
❏ _____
❏ Not sure
❏ Not in a joint environment
❏ Not applicable for other reasons

62. What is the planned length of time a person typically spends in this assignment (in months)? _____

63. How many months do you think this assignment should last?
_____

64. *NOTE: No 64 in nonincumbent version.*

65. Which of the following do you think would be most important to a person in this assignment?
❑ Service core competencies
❑ Prior joint experience
❑ Specialized training and orientation in joint matters
❑ Functional expertise, e.g., operations, intelligence
❑ Other not listed here—please specify:
❑ Not applicable

66. *NOTE: No 66 in nonincumbent version.*

67. A civilian could perform the duties and responsibilities of this position just as effectively.
❑ Strongly agree
❑ Agree
❑ Neither agree nor disagree
❑ Disagree
❑ Strongly disagree
❑ Not applicable

68. *NOTE: No 68 in nonincumbent version.*

69. *NOTE: No 69 in nonincumbent version.*

70. *NOTE: No 70 in nonincumbent version.*

71. Is there anything else you would like to tell us? _____

Bibliography

Bechet, Thomas P., *Strategic Staffing: A Practical Toolkit for Workforce Planning,* New York: American Management Association, 2002 (esp. Chapter 2, "What Is Strategic Staffing, Anyway?").

Bird, Julie, "Air Force to Train Own Navigators (Plans to Pull Out of a Joint Training Program with the Navy)," *Air Force Times,* June 15, 1998.

Booz Allen Hamilton, *Independent Study of Joint Officer Management and Joint Professional Military Education,* McLean, Va., 2003.

Casey, GEN George W., Jr., Army Vice Chief of Staff, speech to the West Point Society and the National Capital, winter luncheon, January 21, 2004.

Center for Strategic & International Studies, *American Military Culture in the Twenty-First Century,* Washington, D.C., 2000.

Chairman of the Joint Chiefs of Staff, *Joint Vision 2010,* 1995.

_____, *National Military Strategy: Shape, Respond, Prepare Now: A Military Strategy for a New Era,* 1997.

_____, *Joint Vision 2020,* 2000.

_____, "Posture Statement," February 8, 2002.

_____, CJCS Instruction 1331.01B, "Manpower and Personnel Actions Involving General and Flag Officers," August 29, 2003b.

_____, CJCS Instruction 1800.01B, "Officer Professional Military Education Policy," August 30, 2004.

Defense Manpower Data Center, *The 1999 Survey of Active Duty Personnel,* Arlington, Va., 2000.

Department of the Air Force, Air Force Instruction 36-2101, "Classifying Military Personnel," April 2001.

_____, *America's Air Force Vision 2020*, 2002.

_____, *Officer Classification*, Air Force Manual 36-2105, April 2003.

Department of the Army, Army Regulation 611-1, "Military Occupational Classification Structure Development and Implementation," September 1997.

_____, "Military Occupational Classification and Structure," Army Pamphlet 611-21, March 1999.

_____, *The Army Vision: Soldiers on Point for the Nation—Persuasive in Peace, Invincible in War*, Office of the Chief of Staff, 2000.

_____, "Serving a Nation at War: A Campaign Quality Army with Joint and Expeditionary Capabilities," www.army.mil/jec/ (as of March 2005).

Department of the Navy, *Naval Operating Concept for Joint Operations*, undated.

_____, *Forward...from the Sea*, 1994.

_____, OPNAV Instruction 1521.2, "Qualification of Navy JOPES Personnel," January 5, 1995.

_____, *Marine Corps Strategy 21*, 2000.

_____, *Manual of Navy Officer Manpower and Personnel Classifications*, NAVPERS 15839I, October 2003a.

_____, *Military Occupational Specialties Manual*, Marine Corps Order P1200.7Y, April 2003b.

Dreyer, Vincent M., Bruce C. Emig, and James T. Sanny, Sr., "The Joint Evaluation Report: Career Enhancer or Kiss of Death," *Joint Force Quarterly*, Vol. 20, Autumn/Winter 1998–1999.

Emmerichs, Robert M., Cheryl Y. Marcum, and Albert A. Robbert, *An Operational Process for Workforce Planning*, Santa Monica, Calif.: RAND Corporation, MR-1684/1-OSD, 2004a.

_____, *An Executive Perspective on Workforce Planning*, Santa Monica, Calif.: RAND Corporation, MR-1684/2-OSD, 2004b.

Evans, Lieutenant Colonel Carl D., *Growing Tomorrow's Leaders in Today's Environment,* Maxwell Air Force Base, Ala.: Air War College, Air University, AU/AWC/RWP094/98-04, 1998.

Feith, Douglas J., "Statement on the Future of NATO by Douglas J. Feith, Under Secretary of Defense for Policy," to the United States Senate Committee on Armed Services, February 28, 2002.

Franks, General Tommy R., testimony before the Senate Armed Services Committee on OIF lessons learned, July 9, 2003.

General Accounting Office, *Joint Officer Development Has Improved, but a Strategic Approach Is Needed,* GAO-03-238, 2002.

_____, *Human Capital: A Guide for Assessing Strategic Training and Development Efforts in the Federal Government,* GAO-03-893G, 2003.

_____, "A Strategic Approach Is Needed to Improve Joint Officer Development," Statement of Derek B. Smith, Director, Defense Capabilities and Management, testimony before the Subcommittee on Total Force, Committee on Armed Services, House of Representatives, March 19, 2003.

_____, *Human Capital: Key Principles for Effective Strategic Workforce Planning,* GAO-04-39, 2004.

Giambastiani, Admiral Edmund P., Jr., USJFCOM, Operational Lessons Learned from Operation Iraqi Freedom, testimony before the House Armed Services Committee, October 2, 2003.

Goldwater-Nichols Department of Defense Reorganization Act of 1986, Public Law 99-433, October 1, 1986.

Goodman, Glenn W., Jr., "Stress the Staffs, Not the Troops—JTASC (Joint Training, Analysis, and Simulation Center) Opening: 'A Giant Step Forward'," *Sea Power,* Vol. 39, No. 2, February 1996.

Graves, Howard D., and Don M. Snider, "Emergence of the Joint Officer," *Joint Force Quarterly,* Vol. 13, Autumn, 1996.

Haraden, Lieutenant Timothy J., "Joint from Day One (Proposal to Combine the Military Academies)," *Proceedings* [U.S. Naval Institute], Vol. 121, No. 7, July, 1995.

Harrell, Margaret C., Harry J. Thie, Jefferson P. Marquis, Kevin Brancato, Roland J. Yardley, Clifford M. Graff II, and Jerry Sollinger, *Outside the*

Fleet: External Requirements for Naval Officers, Santa Monica, Calif.: RAND Corporation, MR-1472-NAVY, 2002.

Harrell, Margaret C., John F. Schank, Harry J. Thie, Clifford M. Graff II, and Paul Steinberg, *How Many Can Be Joint? Supporting Joint Duty Assignments,* Santa Monica, Calif.: RAND Corporation, MR-593-JS, 1996.

Harrell, Margaret C., Harry J. Thie, Peter Schirmer, and Kevin Brancato, *Aligning the Stars: Improvements to General and Flag Officer Management,* Santa Monica, Calif.: RAND Corporation, MR-1712-OSD, 2004.

Joint Chiefs of Staff, "An Evolving Joint Perspective: US Joint Warfare and Crisis Resolution in the 21st Century," white paper, Joint Vision and Transformation Division, January 28, 2003.

Joint Staff, *The Future Joint Force: An Evolving Perspective,* JROCM 022-03, January 28, 2003.

_____, "Cumulative Joint Duty Credit for Combined Joint Task Force Headquarters Assignments in Approved Operations," memorandum, J-1A 00165-04, July 12, 2004.

Johnson, David E., "Preparing Potential Senior Army Leaders for the Future: An Assessment of Leader Development Efforts in the Post–Cold War Era," Santa Monica, Calif.: RAND Corporation, IP-224-A, 2002

LINK Perspective (Career Magazine for the Navy Professional), 2003. Online at www.npc.navy.mil/ReferenceLibrary/Publications/LinkPerspective/ LINK-Perspective+Archives.htm (as of March 2005).

Locher, James R. III, "Taking Stock of Goldwater-Nichols," *Joint Forces Quarterly,* Autumn 1996.

_____, "Has It Worked? The Goldwater-Nichols Reorganization Act," *Naval War College Review,* Vol. 54, No. 4, Autumn 2001.

MacGregor, Douglas A., "The Joint Force: A Decade, No Progress," *Joint Force Quarterly,* Vol. 7, Autumn 2001.

Mahnken, Thomas G., and James R. FitzSimonds, *The Limits of Transformation: Officer Attitudes Toward the Revolution in Military Affairs,* Newport, R.I.: Center for Naval Warfare Studies, 2003.

Martin, Joann, *Organizational Culture: Mapping the Terrain,* Thousand Oaks, Calif.: Sage, 2002.

National Defense Authorization Act for Fiscal Year 2002, Public Law 107-107, 521-529, December 28, 2001.

National Guard Bureau, "2004 National Guard Posture Statement," www.ngb.army.mil/ll/04posture/ (as of February 2005).

Noonan, Michael P., and Mark R. Lewis, "Conquering the Elements: Thoughts on Joint Force (Re)Organization," *Parameters (US Army War College Quarterly)*, Vol. 33, No. 3, Autumn 2003.

Ochmanek, David, *NATO's Future: Implications for U.S. Military Capabilities and Posture*, Santa Monica, Calif.: RAND Corporation, MR-1162-AF, 2000.

Paauwe, J., and P. Boselie, *Challenging (Strategic) Human Resource Management Theory: Integration of Resource-Based Approaches and New Institutionalism*, Rotterdam, The Netherlands: Erasmus Institute of Management, ERS-2002-40-ORG, April 2002.

Quigley, Commander John M., "Creating Joint Warfighters," *Proceedings* [U.S. Naval Institute], Vol. 121, No. 9, September 1995.

Rosenberger, Lieutenant Colonel John D., "Key to Joint Readiness," *Proceedings* [U.S. Naval Institute], Vol. 121, No. 9, September 1995.

Rumsfeld, Donald H., testimony before the Senate Armed Services Committee on OIF lessons learned, July 9, 2003.

Scales, Major General (Ret.) Robert L., "Building on Brainpower: Developing a Community of Lifelong Learners Is Key to Transforming the Military," *Armed Forces Journal International*, Vol. 140, No. 7, February 2003.

Schank, John F., Harry J. Thie, and Margaret C. Harrell, *Identifying and Supporting Joint Duty Assignments: Executive Summary*, Santa Monica, Calif.: RAND Corporation, MR-622-JS, 1996.

Schank, John F., Harry J. Thie, Jennifer Kawata, Margaret C. Harrell, Clifford M. Graf II, and Paul Steinberg, *Who Is Joint? Reevaluating the Joint Duty Assignment List*, Santa Monica, Calif.: RAND Corporation, MR-574-JS, 1996.

Scully, Megan, "Joint but Not Yet Seamless: Military Exercise Exposes Cultural Gaps in U.S. Inter-Service Operations," *Defense News*, February 2, 2004.

Secretary of Defense, "Legislative Priorities for Fiscal Year 2004," Memorandum to CJCS Secretaries of the military departments, Under SecDef, Dir Defense Research and Engineering, Asst SecDef, Gen Counsel of the DoD, IG of the DoD, Director Ops T&E, Assts to the SecDef, Dir Admin and Mgt, Dir Net Assessment, Dirs of the Defense Agencies, Dirs of the DoD, Field Activities, September 17, 2002.

_____, *Annual Report to the President and the Congress*, 2003.

_____, "Limit on Good of the Service Waiver Policy," memorandum to secretaries of the military departments, August 27, 2003.

Strom Thurmond National Defense Authorization Act for Fiscal Year 1999, Public Law 105-261, October 17, 1998.

Thie, Harry J., and Roger A. Brown, *Future Career Management Systems for U.S. Military Officers,* MR-470-OSD, Santa Monica, Calif.: RAND Corporation, 1994.

Thie, Harry J., Margaret C. Harrell, and Robert M. Emmerichs, *Interagency and International Assignments and Officer Career Management,* Santa Monica, Calif.: RAND Corporation, MR-1116-OSD, 1999.

Under Secretary of Defense for Personnel and Readiness, *The Army Vision: Soldiers on Point for the Nation Persuasive in Peace, Invincible in War,* 1999.

_____, *Military Personnel Human Resources Strategic Plan: Change 1*, 2002.

U.S. Department of Defense, "Revolutionizing Military Personnel and Pay," DIMHRS fact sheet, Office of the Under Secretary of Defense for Personnel and Readiness, undated.

_____, *Audit of the Joint Operation Planning and Execution System*, Officer of the Inspector General, Report 94-160, June 30, 1994.

_____, DoD Instruction 1300.20, "DoD Joint Officer Management Program Procedures," December 20, 1996.

_____, *Quadrennial Defense Review Report,* September 30, 2001.

United States Code, Armed Forces, Subtitle A, General Military Law, Title 10, Part II, Personnel, Chapter 38, Sec. 667, *Annual Report to Congress*.

U.S. Coast Guard, *Coast Guard 2020: Ready Today…Preparing for Tomorrow,* 1998.

The White House, *The National Security Strategy of the United States of America*, 2002.